THE
RAW
FOOD
Diet Myth

THE RAW FOOD

Diet Myth

What You Should Know about the Raw and Living
Food Lifestyle to Improve Your Health, Fitness, and Life

Ruthann Russo, PhD, MPH

DJ Iber Publishing, Inc
Bethlehem, PA

The Raw Food Diet Myth, by Ruthann Russo

Published by:
DJ Iber Publishing, Inc.
One Bethlehem Plaza, Suite 1010
Bethlehem, PA 18018
www.djiber.com

ISBN–13: 978–0–9799061–2–1
Library of Congress Control Number: 2008931273

Editor: Emmalea C. Russo
Copyeditor: Ginger McQueen
Proofreader: Gary Warren
Cover and Interior Design: Desktop Miracles, Inc.
Cover Art: Emmalea C. Russo

Printed in the United States of America

40535865 6/09

Advice given is general. Neither the author nor the publisher is engaged in providing medical, health, or legal services. Readers should consult professional counsel for specific questions. The author and publisher expressly disclaim responsibility for any adverse effects arising from the use or application of the information contained in this book.

The author and editors have made every effort to verify the accuracy of all the resources, including Web sites, referred to in this book. However, organizations and Web sites mentioned may change or cease to operate after this book is published. Visit www.ruthannrusso.com and www.djiber.com for current resources.

Contents

Appendices

· · · · · · · · · · · · · · · · · · ·

Introduction and Overview of
Raw and Living Food Principles

The myth is that raw food is not just a diet, it's a lifestyle. Raw food is also a revolutionary philosophy. A revolution is a fundamental change in the way of thinking about something.[1] The raw food movement changes the way we look at food; therefore, it is a revolution. Philosophy is an interpretation of the way things fit together.[2] The raw food movement looks at the way food, living, our treatment of the earth, our treatment of each other, and our quest for physical, spiritual, and mental health all fit together. Therefore, it is also a philosophy. The purpose of this book is to introduce you to the revolutionary philosophy of raw and living foods by pulling together all of the components, including diet, into one place. The book provides you with information to make a conscious decision about whether you will incorporate any of the raw and living food philosophy into your own philosophy of living.

Raw versus Living Foods

The terms *raw* and *living* are often grouped together when describing foods. And in some places in the book I use them interchangeably. It is, however, essential to clarify the difference between the two terms. *Living* foods are those which still have live enzymes circulating in them, the more, the better. Examples of *living* food include uncooked broccoli, romaine lettuce, and apples. Within the context of this book, *raw* foods are any foods that are *grown* and then eaten in their uncooked form. If it wasn't grown in the ground (on a plant, a tree, a vine or as a root), it's not raw or living. You can

1

eat fish, meat, and even milk and cheese products raw. These foods, however, are part of a diet that includes animal products and are not addressed in this book. Examples of *raw* foods that grow in the ground include raw almonds, pumpkin seeds, and lentil beans. You can revitalize these foods by soaking them, which causes them to sprout. During this time, they begin to produce enzymes again.

While all living foods are raw, not all raw foods are living. It is possible to revitalize many raw foods through the sprouting process. However, not all raw foods can be sprouted. Examples of raw foods that cannot be revitalized through sprouting include raw cacao beans (from which chocolate is made), sweeteners like agave, maca, mesquite, and dried herbs. Degree of life, as we will see, is a point of contention within the raw and living food movement. But if you understand this and the differences inherent in living versus living *and* raw foods, you will be able to make some good decisions regarding your own nutrition.

The Definition of "Uncooked"

The raw food movement is a return to natural, almost biblical practices. For example, in the *Essene Gospel of Peace*, Jesus is quoted as saying, "Live only by the fire of life, and prepare not your foods with the fire of death, which kills your foods, your bodies, and your souls also."[3] The philosophy prescribes eating only nonanimal food, and only when it is still in its *living* or *raw* state. This means the food is organic, unprocessed, unpasteurized, and unpreserved. In addition, the food must be uncooked to ensure enzymes in the food remain alive (more on enzymes later). Because most enzymes can survive heat of up to about 115° Fahrenheit, raw or living foodists heat food up to 115° F. This methodology would also be consistent with the way that the Essenes *baked* their breads in the sun over 2,000 years ago. Today, we use dehydrators to heat foods to 115° F since most ovens cannot be set below about 175° F. Dehydrators work by gently heating the air and blowing it throughout the food-drying area. Using no heat sources would be ideal for enzyme retention. But dehydrating adds variety to the raw and living food diet. Instead of just fruits and vegetables, you can use the dehydrator to make foods like flatbread, dried fruit, and crackers to spice up your meals.

Enzymes Can Improve Your Life

Eating only foods that are uncooked or heated up to 115° F allows the food to retain enzymes that are destroyed in the normal cooking process, which for most of us is 350 to 425° F. Enzymes are essential for digestion. If we allow the foods we eat to retain their living enzymes, then our bodies rely on those enzymes, making our own digestion process more efficient. Essentially, eating raw and living foods allows our body to use its energy to stay healthy and not produce as many enzymes. We also free our bodies from digesting the unnatural and sometimes toxic substances contained in cooked, processed, genetically modified, or pesticide-infested foods when we eliminate these kinds of foods from our diets. Living foods also have a high level of energy that translates into higher energy levels for us when we eat them. Other components of the philosophy include treating everyone and everything with respect, including ourselves, the food we eat, the earth, and others.

Every revolution, including the raw and living food movement, has its leaders. It is difficult, however, to pin down just one leader. Dr. Norman W. Walker published and taught *vibrant health*, his version of raw foodism, from the early 1900s until his death in 1985. And Dr. Ann Wigmore, published and taught *natural living*, her version of raw foodism from the 1930s until her death in 1994. In 1960, Viktoras Kulvinskas started the Hippocrates Health Institute with Dr. Wigmore and continued her work. Since then, many leaders, each an expert in one or more spheres of raw and living food philosophy, have surfaced. Gabriel Cousens, MD, MD(H), DD, has developed and shared his philosophies on spirituality and nutrition. David Wolfe has pioneered the way for Americans to obtain easy access to high-quality living foods. Victoria Boutenko has passed on to us her approach to transitioning your entire family from a processed diet to a raw diet. Joel Fuhrman, MD, has conveyed the mainstream physician's approach to fasting, which is an important component of the raw and living lifestyle. The extent to which the raw and living food philosophy has been adopted by individuals around the world is due in large part to the open and collaborative efforts of these leaders.

For the first 20 years of my life, I was a meat eater. Over the past 25 years, I transitioned from being a vegetarian to a vegan who ate about 50 percent cooked foods. Two years ago, the transition to a raw diet was a relatively easy one for me. It was a logical next step, and I was motivated to find a *nutritional* solution that might help prevent my daughter's seizures. We were looking for something that could either replace or supplement the prescription medication she was currently taking. When we decided to try raw food on the advice of a physician and a nutritionist, transitioning as a family made the process easier. The one thing that made the transition effortless for me was the fact that chocolate, in its raw form, is an acceptable raw food.

The standard American diet, also known by the acronym SAD, is a diet that is high in animal fats, high in unhealthy fats, high in processed foods, low in fiber, low in complex carbohydrates, and low in plant-based foods.[4] The transition from a SAD to consuming mostly or all raw and living foods can be both physically and emotionally challenging. Physically, your body will be reacting to the elimination of toxins found in processed foods and sugars. Emotionally, you will experience withdrawal from the habits and food dependencies you have built up over the years. Therefore, if you do make the decision to change, it is important to move gradually to a raw food diet. Your goal should be to trigger the cleansing action of a raw food diet with a minimal amount of dietary change necessary to see results.[5] You can use the assessment in nutritionist Natalia Rose's book, *The Raw Food Detox Diet,* as a guide. In the book, Rose categorizes everyone into a transition group from 1 to 5 based upon your answers to questions about your historical and current eating habits, health practices, and age. Then, depending on your group, she recommends the types of foods that are likely to work best for you.[6]

Dr. T. Colin Campbell, a professor at Cornell University and author of *The China Study,* published in 2007, provides some of the most convincing and scientific support for a vegan or "plant-based diet."[7] Protein can be found in leafy green vegetables like spinach, kale, and collard greens. It can also be found in high amounts in sea vegetables like dulse, nori, and kelp. Dr. Campbell spent the last 25 years conducting studies for Cornell and the National Institutes of Health (NIH) on the role of protein and plant-based foods in diet. He and his research team found, among other things, that increasing the amount of plant-based protein in your diet can actually counteract both genetic predisposition to disease and exposure to toxic substances. In an interview in *Living Nutrition* journal, Dr. Campbell stated that "organic and raw is the most natural way to go [in your dietary habits]—it is the ideal."[8]

Raw Food Trends

In my travels through the raw food circles, I found several trends that appeared to be common among all raw foodists whom I communicated with or observed. What I found was that the raw food or living food philosophy is as follows.

Integrative

The raw food diet myth is just that—that the raw food diet is just a diet. In fact, the raw and living food philosophy goes far beyond food. It integrates all areas of our lives, making it not just a diet, but a lifestyle. The different leaders of the raw food movement have brought this philosophy to one that encompasses, in addition to food preparation, virtually all areas of life: our attitude, how we treat people, how we treat the earth, our fitness, and even our spirituality.

Good for You

I am a skeptic by nature. I spend a lot of time researching, comparing different viewpoints and outcomes, and I look specifically for the negative impact of anything. I have yet to find anything unhealthy or "bad" about the raw food diet. When you transition from a SAD to raw foods, your body's adjustment to the elimination of toxins that have built up over time, especially through a change in your elimination habits, can be unpleasant. But, as with any significant change, an initial imbalance followed by a return to homeostasis is to be expected.

A Paradox that Embraces Both Abundance and Fasting

Instructors at the raw retreats I have attended frequently used the term *abundance* when talking about the dish we were about to prepare or the meal we were about to eat. While *abundance* was never defined by any of them, their focus was on all of the goodness that the earth gives to us. Most raw food dinners I have attended are rich in types of foods, color, texture, as well as the overall amount of food. The term *abundance* can refer not only to amount, but also to the characteristics of the food itself. In addition, there are many who posit that on the raw food diet, an individual can eat to his or her heart's content. There is some disagreement about this in the raw food community. Here's one way to look at it: 5,000 calories of raw food is 1,000 percent better for you than 2,000 calories of the standard American diet.

Fasting is embraced by every leader, to a certain extent, in the raw food community. Fasting is generally seen as a way of giving the body a break from digestive stresses. Fasting also can have mental and emotional benefits that can be as simple as noticing your orientation toward food and, at least for a day or so, resisting those urges. There are some differences in terms of the length of a fast (one day to 50 days) and the nutritional content of a fast (fruit juices, vegetable juices, or water). Cousens defines fasting as, "the elimination of physical, emotional, and mental toxins from our mind and body, rather than simply cutting down on or stopping food intake."[9]

Cousens also claims that fasting is the ultimate way to reactivate what he refers to as the *youthing* gene. Cousens also says, "On the physiological level, fasting works by rapidly removing dead and dying cells and toxins. But fasting also stimulates the building of new cells. Aging occurs when we have more cells die than are being built. *Youthing* happens when more new cells are produced than are dying."[10]

While fasting is viewed predominantly in terms of its relationship to food, you can fast from activities as well. Examples include fasting from media, shopping, gossip, or other activities that interfere with your ability to feel happy and alive. Regardless of what you are fasting from, one of the primary purposes of a fast is to increase your ability to be aware, conscious, and mindful.

One cautionary statement about fasting. Most spiritual leaders advocate fasting. Mahatma Ghandi is probably the leader most well-known for his use of fasting as a protest and to help him endure his imprisonment. However, if you have a chronic illness, most physicians suggest engaging in a fast only with medical supervision. There are several retreats including those at Cousens' Tree of Life and Clement's Hippocrates Institute that provide closely supervised fasts for up to 21 days.

Cultlike, But Not a Fad

Just as with anything not considered to be in the mainstream, there is a cultlike quality to the raw food philosophy. A cult can be thought of as religious in nature. However, a cult is also a group of people with a great devotion to an idea or movement that is considered by nonmembers to be a fad. If your definition of raw food is limited to the dietary component only and you have a limited understanding of the science behind it, you may view the raw food movement as a fad. However, it is likely that as subscribers to the raw food philosophy increase in number and continue their devotion to all components of the lifestyle, not just diet, the view of it as a fad will fade.

Entrepreneurial

The raw food movement is wrought with entrepreneurs. Most of them are really smart people who are willing to share their findings with anyone willing to listen. With the exception of Web-based resources like Wikipedia, the sharing of information openly with no expectation of payment in return is a foreign notion to most of us, especially in the United States. While you will have to pay to attend comprehensive workshops or buy their books, you can piece together a solid plan just from the resources that are offered for free by raw and living food entrepreneurs. Two good examples of raw food entrepreneurs who fit this description are David Wolfe and Gabriel Cousens. Wolfe is a former attorney who now owns and manages the Sunfood store and Web site to supply us with a wide variety of raw materials to prepare raw dishes. Wolfe has six different Web sites where you can obtain free information on his philosophy and experience with raw food. Gabriel Cousens, another entrepreneur, is a medical doctor and doctor of divinity who offers a free newsletter, free Webcasts and speaks to many groups. These educated entrepreneurs with connections to the mainstream are important components of a new movement because they bring credibility and sustainability to it.

Universal

Most of us subscribe to pieces of the raw food philosophy. We have to. It is basic and natural. As you piece through the different components of the philosophy, ask yourself which ones you already embrace. Eating organic, fresh, uncooked fruits and vegetables? Having respect for the earth? Respecting others? Being grateful for something each day? Believing that there is some being, bigger than any of us, guiding our existence? Appreciating and loving yourself so you can generate positive energy to others? The raw and living food philosophy and lifestyle addresses concepts as small as atoms to as large as the universe and pretty much everything in between.

Supported by Converts

Ninety-five percent of individuals who consider themselves to be *raw foodists* are converts. Although this statement is not supported by scientific research, it is supported by many of the individuals I interviewed. Everyone has a story about *converting* to raw food. Many were driven by a health-related cause. For example, Arnold Kauffman, owner of Arnold's Way, a raw food café in Lansdale, Pennsylvania, was motivated to try a raw food lifestyle after a near heart attack. Ann Wigmore was motivated by a cancer diagnosis. The Boutenkos were motivated by Sergei's diagnosis of diabetes. Dr. Joel Fuhrman was motivated by an injury he could not overcome.

As far as I can tell, there is no adult living today who was born a raw foodist.

And in my many conversations with individuals about their transition to the raw food lifestyle, almost everyone started with a focus on food. Eventually, that focus began to encompass other components of their lives including their spirituality, emotional health, occupation, and relationships. The raw food movement is a *philosophy* and not just a diet. It is a way of life.

Supported by Science

There is a solid scientific basis for the raw food philosophy. In order to be an informed raw foodist, you need to understand the basic science of nutrition. You do not need a degree in nutrition, but it is helpful if you understand the basic concepts related to body pH, food combining, production of enzymes, and cellular regeneration. More importantly, when you understand the simple science of food and your body, it is easier for you to arrive at good lifestyle decisions.

The fact that the raw food philosophy is supported by science differentiates it from diets that are generally not sustainable. While subscribers to the raw food philosophy do not have to know significant details about the scientific basis for the diet, it is comforting that most instructors and food preparers are very aware of the science behind the food plan. I have experienced, on numerous occasions, raw food instructors flawlessly describing the property differences between flax, olive, and hemp-seed oil. Most can usually describe detailed facts such as that the human body can not directly process flaxseeds, so the seeds need to be ground or modified in some other way during the food preparation process in order for us to obtain maximum nutritional benefit.

Not Exempt from Junk-Food Junkies

Even raw foodists go astray. I am a good example of this. I have already mentioned my love of chocolate and the fact that raw chocolate made from raw cacao beans and agave syrup (a sweetener obtained from cactuslike plants) is one of my favorite treats. On days when I am under enormous stress, I have had nothing but green juice and raw chocolate to eat. The good news is that you can recover from this type of digression much more quickly than if you are partaking of traditional chocolate candy.

Yes, you can gain weight on a raw food diet. This was discussed during the lectures I heard at a raw food retreat, and many people attested to the fact. You will never become obese on a purely raw food diet, but you can gain weight if you intake a high percentage of calories from seeds and nuts. Raw seeds and nuts, like pumpkin seeds (the highest in protein of all seeds and nuts), sunflower seeds, almonds, and walnuts are considered by many to be the meat replacement in the raw foodists' diet. They are also eaten more frequently during the cold months. It's wise to eat seeds and nuts in moderation.

Supported by a Spiritual/Religious Component

As I noted previously, leaders like Gabriel Cousens, who wrote *Spiritual Nutrition*, espouse a strong connection between the practice of eating and spirituality. Specifically, Cousens states that "what we eat affects the quality of the functioning of our mind, whether our mind is noisy, quiet, at peace, or irritated." He also says, "Our food choices reflect our state of harmony with ourselves, the world, all of creation."[11] Cousens goes on to describe the fact that viewing nutrition in this manner helps us all to live a harmonious and peaceful life on earth.

A Conscious Decision

On the raw food diet, you become conscious of what you are taking into your body. This is partly because you need to assess the quality of the food—it could be as simple as determining whether the vegetables you are buying are *conventional* or *organic*—and partly because you feel healthier and more alive as a result of your nutritional habits. The different textures, colors, and tastes are also likely to pique your levels of awareness. Consciousness goes far beyond what you eat—but for most of us, that's the starting point. As your eating habits change, you begin to become more aware of other aspects of your life. You will be more conscious and mindful of your treatment of others, your treatment of yourself, and every activity you engage in.

About Healthy Conflict

As with any philosophy, there exist some areas of conflict amongst the raw food leaders. I call it healthy conflict because, from what I have seen, each leader is aware of and acknowledges the different beliefs her peers have. They do this with respect for each others' opinions and foundations of thought. Each has strong convictions about his beliefs and each has spent significant time amassing information to support his theory. A few examples include Gabriel Cousens' support for using salt in all raw food preparation, while Brian Clement has eliminated the use of salt altogether. Natalia Rose recommends drinking green juices daily (juices have all fiber removed), while Victoria Boutenko recommends drinking green smoothies, which retain all the fiber. It is important to be aware of these differences and the basis for each so you can try different options and make your own decisions.

Colorful

Both the food and the people who are raw foodists have a strong colorful glow to them. A recent raw food meal that I had at Sarma Melngailis' restaurant, Pure Food and Wine in New York City, contained the following natural colors: green (marinated kale salad), red (roasted red pepper—but not heated above 115° F!), purple (black olive compote), yellow (nut cheese), and orange (butternut squash soup and raw pumpkin pie). It was almost as enjoyable to view the food as it was to eat it!

Not too long ago, my daughter, Emmalea, showed me two pictures of myself taken at different times. She described the photos as a before (pre-raw) and after (post-raw) pictures. My weight is the same in both snapshots. However, in the pre-raw picture, my face is almost pure white and my eyes are a little swollen and puffy (probably *carb-face*). However, in the post-raw picture, my face has no puffiness and my color is a rosy orange, like a soft suntan. I have not sunbathed for over 20 years. I attribute the change in color totally to my diet.

Talked About in Percentages

"I am 75 percent raw, what percentage are you?" This is one of the most common questions you will be asked in a raw café, raw retreat, or raw retail store. When I first began to explore the lifestyle, I was not sure what it meant, and I am still not sure how anyone is truly capable of calculating their dietary intake with such precision. At a recent Tree of Life retreat that I attended, Gabriel Cousens taught that eating 100 percent live foods is the most powerful way to elevate the spiritual aspects of health but over a two year period, an 80 percent live food diet (and lifestyle) will give all the health benefits of eating 100 percent live foods. Any degree of food intake that is raw is beneficial. And if that raw food is organic, even better. Most raw foodists take proactive steps to ensure their food is organic. Organic and raw foods go hand in hand.

Serene and Peaceful

During the past two years, I have met several people who subscribe to the raw food philosophy. I have observed them to be completely and truly serene. You can spot them. It's in their coloring, their nature, their smile, their relaxed and confident way of going about whatever it is that they are doing. This is particularly true of people who are involved with raw food preparation for a living. Sarma, the owner of Pure Food and Wine in New York City, is one of them. The owner of Pastel Luna, a raw food café in Harrisburg, Pennsylvania, is another. And, the entire kitchen staff at the Tree of Life in Arizona display raw food serenity. I have consistently observed this peacefulness, but I am unable to attribute it to any other common cause except the raw and living food lifestyle.

Thus far, the observations that I have shared about raw food philosophy have been predominantly positive. It is also important to address improvements that can be made in the raw food practice and philosophy. These improvements include the following:

Demand Outweighs Supply

Currently, there is a limited population with the knowledge of raw food preparation. In order for raw food to be enjoyed by the masses on a larger scale, the supply of trained people needs to increase. The shortage of freshly prepared raw foods, especially in metropolitan areas, is apparent. For example, one evening I was dining at a raw food restaurant and by 9 p.m. two of the entrees had already sold out. This is something that rarely happens in traditional restaurants, especially in New York City.

Lack of Convenience

Unlike traditional American convenience foods, raw food is not generally available just anywhere. Even finding a good salad made with high-nutrient lettuce, not iceberg lettuce, is fairly challenging if you are traveling or in an airport. Before I travel, I Google the city or area I am visiting for any raw food cafés. Seventy-five percent of the time, I can find a raw resource—and about half of these resources offer more than juice and salad. Most of these places are not conveniently located directly off a major highway. It does take a little more effort to support a raw food lifestyle, especially when traveling, but it can be done. There are a myriad of Web sites that sell raw food treats and bars, most of which can be packed for traveling. These Web sites are listed in an appendix at the end of this book.

Food Preparation Can Take Time—Lots of It!

There is no doubt about it; other than salads and fresh fruit, the majority of raw food preparation takes a lot of time to prepare. If you had to live a life confined to eating only salads and fruits, chances are great that you would abandon the raw lifestyle very quickly. Luckily, the leaders and their students have gone to great lengths to devise creative ways to prepare raw foods that are tasty, are beautiful to look at, and still retain all of their natural nutrients. The downside to this is that the preparation can take many hours if the food needs to be dehydrated. The dehydrated foods add variety and texture to a raw food meal. Luckily, there are some shortcuts, and many are described in uncook books like *Rawvolution* by Matt Masden and *Raw Food Made Easy for 1 or 2 People* by Jennifer Cornbleet.

Requires Self-Experimentation

Self-experimentation in nutrition and diet is important for everyone for several reasons. First, it makes you aware of your food intake. Second, it makes you more aware of your physical and emotional reaction to certain foods (good and bad). Third, any positive change in your dietary habits, even slight, is likely to give your body a "jump start" and, depending on your age and other circumstances, it may even boost your metabolism a bit.

Prepared Raw Foods Can Be Expensive

Foods at most raw restaurants are highly priced. In New York City or San Francisco, dinner entrees are priced competitively, and sometimes even higher than traditional restaurants that serve meat. The reason for this is twofold. First, the resources needed to prepare the food are often greater than in a conventional restaurant. Second, high-quality organic nuts, which serve as the basis for most dinner entrees in raw restaurants, are expensive and can cost as much as 15 dollars per pound. Desserts are also very pricey, including ice cream. Raw vegan ice cream, which is made from young Thai coconut meat, cashews, and agave, is one of the most delicious desserts ever. However, because of the time and expense required to make it, raw ice cream is difficult to find and when you do find it, it can cost up to 20 dollars a quart. My husband, Joe, will go out of his way to stop at the Pure Food and Wine restaurant just to order their classic sundae because he, a non-raw foodist, believes it is the best dessert he has ever tasted.

Your Body Goes through Withdrawal

I have mentioned previously that the transition from non-raw to raw foods is one that should be undertaken with care. Natalia Rose's questionnaire in her book *The Raw Food Detox Diet* will help you rate yourself, and then you can follow her recommendations for foods to eat and how to transition properly for your level. Withdrawal from any toxin is not fun. When you get to the other side, you are much better off, but getting there can be a difficult journey.

Causes a Significant Change in Elimination Habits

This issue goes hand in hand with the above statement about withdrawal. Initially, withdrawal will impact your gastrointestinal system. You may have slight stomachaches or nausea followed by frequent elimination. You will need to manage your lifestyle around these changes during the initial withdrawal. Once you have gotten to a plateau, which can take one to four weeks, depending on the level of toxins from processed, nonorganic, pasteurized foods in your system, you can count on consistency as long as you do not drastically change your habits.

How to Use This Book

Use this book first to learn what the raw and living food lifestyle is all about. I have tried to describe all of the different components of the raw and living food lifestyle. Then, use this information to determine what parts of the philosophy, if any, you can use in your own nutritional practices and in your life.

Note on Medical Advice

There are different places in the book where I describe the medical advice given by different physicians involved in the raw food movement. These are pieces of information meant for illustrative purposes only. The information is not meant to be specific medical advice to be acted upon by anyone.

THE *basics*

· · · · · · · · · · · · · · · ·

CHAPTER 1

Raw Stories

· ·

March 20, 2007 was the date I first heard the term *raw food diet.*
Standing in Borders Books in the Time Warner Center in New York City, I
was beginning to feel that I would never find what I had been investigating
for months online and in every bookstore I came across: a nutritional philosophy that could help neurological health, in particular, epilepsy. My daughter,
Emmalea, was diagnosed with juvenile myoclonic epilepsy the previous year
and the numbers and types of antiepileptic drugs seemed endless, not to mention potentially dangerous and ineffective. So, nutrition became a primary
avenue for possible solutions.

I was in the diet and nutrition aisle at Borders when I noticed a book
by nutritionist Natalia Rose, *Raw Food: Life Force Energy.* I picked it up more
because the dust jacket was very attractive, and it showed a picture of Rose
walking confidently past an outdoor market stocked with fresh fruits and
vegetables. While I did not think this book was the solution, I was intrigued

by Rose's discussion of light, energy, and the vibration principle. I bought it, as well as her first book, *The Raw Food Detox Diet,* and I began to read.

Rose provides an excellent description of the raw and living food diet, yet I was confused about how this lifestyle could be possible, let alone enjoyable. Then I noticed Rose's book listed raw food resources. One of those resources was Arnold's Way, a raw food café, located about 40 miles from my home in Pennsylvania. When I first described the diet and the idea of visiting Arnold's Way to Emmalea, she was up for it. That Sunday, we made the trek to Lansdale to meet Arnold and see the café. Before we left the café that day, we received some spontaneous instruction in both raw food nutrition and preparation from Arnold Kauffman, the owner. The café was full of food preparation supplies, raw ingredients, fresh, local organic produce, and Arnold's own smoothies and prepared dishes. Emmalea and I had our first raw meals that night. She had banana ice cream and raw berry pie made with ground raw nuts for the crust and dates for sweetener. I had raw pizza, made with flaxseed crust, blended sun-dried and fresh tomato sauce, and fresh chopped yellow peppers and onions as the topping. We fell in love first with raw food and then with the lifestyle. Our lives would never be the same.

Because I had been a vegetarian and then a vegan for almost 20 years, the transition to vegan raw food was not very difficult for me. It was actually a welcome change—one that increased my energy level and ability to manage stress. Each individual is unique, so the response to raw food will not be the same for everyone. Search for what works best for you. The table below gives you an idea of typical meals for me.

Everyone has a story about how they found raw food. As I have traveled around the country and spoken to many raw foodists, I have enjoyed hearing their stories. And, of course, I have enjoyed reading about the stories of the raw food leaders and authors. The first story I heard about raw food transition was Arnold Kauffman's, owner of Arnold's Way Cafe. In his 40s, Arnold suffered a near heart attack and was told that he could not leave the hospital without bypass surgery. Never one fond of the medical establishment or the thought of surgery, Arnold signed himself out of the hospital. Arnold spent over four weeks on a water-only fast, learned the raw and living diet,

Table 1.1–Typical Foods in My Raw Diet

BREAKFAST:	Green juice—20–30 oz made with romaine lettuce, spinach, kale, celery, or collard greens (whatever greens are in the fridge), lemons, and ginger root.
LUNCH:	One or two of the following: green salad, raw veggie salad, raw soup, raw pizza, raw flaxseed crackers. Salad dressing is either raw olive oil and lemon juice or, occasionally, raw honey mustard.
DINNER:	One or two of the following: nut meat patties, veggie salad, raw taco salad with nut meat, zucchini hummus and crackers, raw nut "cheese" and veggies, zucchini spaghetti.
SNACKS:	Raw nut mixes, raw chocolate, goji berries, mesquite truffles, raw ice cream (made with young Thai coconut meat, whenever I can find it), raw pecan pie, sprouted buckwheat cereal with fresh almond milk.
DRINKS:	Water, green juice, organic herbal tea
NOTE:	Fruit does not work well for me. People who are prone to candida infections should stay away from excess fruits and starches and I am one of them. But fruit works well for my husband Joe, who eats a diet that is not all raw. His daily regimen includes a "Green Smoothie"—bananas and whole greens (kale, spinach, collard greens) that he blends in our Vita-Mix blender. He also eats cooked fish dinners once or twice a week. This is a process that he has worked out and creates balance for him. As a result of adopting this specific approach to his diet, Joe has lost about 20 pounds, looks better, and feels much more energetic.

and became *a new man.* The science behind fasting, although it has never been proven, claims that when the body is deprived of food, it must become as efficient as possible. So to increase efficiency when food-deprived, as when you are fasting, the body sheds sick and old cells. Shedding sick and old cells results, supposedly, in a natural healing and rejuvenation process.

Arnold has taken this gift he was given, a naturally healthy life, and used it to help others. If you visit his store, Arnold's Way, he will gladly share with you photo albums filled with pictures of customers he has helped to achieve the spontaneous healing process through juicing, smoothies, and fasting. During my many trips back to his store, I have personally met customers who attribute their good health to Arnold. These include customers who have or previously had multiple sclerosis, prostate cancer, heart disease, arthritis, lupus, and even vaccination reactions. This is a small group of people, and while Arnold's story would probably not hold weight in the scientific community, it's certainly worthy of being called at a minimum, a case study with a positive outcome— perhaps justifying additional research to see if these practices can indeed result in the same positive outcomes for others.

The Tree of Life (TOL)

My exploration of the raw and living lifestyle led me to Gabriel Cousens' Tree of Life Rejuvenation Center, where raw food preparation courses are taught. While at the TOL, I had the opportunity to speak with many employees and other students. There were 14 students in the class. Three of those students were insulin-dependent diabetics who were also enrolled in a fasting and raw food program designed to help them control their insulin levels with the hope of eventually weaning off of insulin injections completely. Were this to occur, it would be monumental because individuals with insulin-dependent diabetes traditionally have lived their entire lives dependent on insulin with no hope for eliminating that dependency. One of the diabetes program participants, who had been diagnosed with diabetes for less than a year, was off of insulin completely by day 21. Another program participant, who had been insulin dependent for 22 years, was able to decrease her insulin to 33 percent of the amount she had taken for the last decade by day 30. I do not know a lot of other facts about these specific participants. What I do know is that the hope and vigor this possibility produced in these individuals created a positive wave that spread throughout the TOL. In 2008, Dr. Cousens published *There is a Cure for Diabetes: The Tree of Life 21-Day+ Program,* which addresses stories like these in detail and is probably worth a read for anyone with diabetes or for the loved one of someone with diabetes.

A unique story about the transition to the raw lifestyle was related to me by a man I will call Jim. At the age of 65, Jim was a contented, peaceful resident of the TOL, but this transition did not come about because of health issues, the way that raw food grabs most people. Jim spent all of his professional life as an airplane pilot. He was stressed, had a bad diet, and was getting ready to retire.

He was in good health, but he felt something was missing in his life. Jim found that missing puzzle piece when he met and studied with Gabriel Cousens. In addition to being a medical doctor, Dr. Cousens is also a doctor of divinity who received rabbinical initiation. As a result, the TOL engenders a very strong spiritual community. Jim realized that spirituality was the missing component in his life. After meeting Dr. Cousens and being introduced to the spiritual component of the raw and living food lifestyle, at least as it is practiced at the TOL, Jim made the decision to dedicate his life to helping others at the TOL and spreading the raw and living food lifestyle and philosophy.

Natalia Rose

Natalia Rose's story may be the best raw food story simply because it is not a story. Rose is a young nutritionist who trained at New York University. Through her studies and assessments of other nutritional alternatives, she chose what could be termed a modified raw food diet. Rose probably represents the new, emerging generation of raw foodists—those who don't have much of a story, but just eat this way because it is the healthiest way they have found to sustain their bodies. Rose practices and describes a modified raw food diet in her books. Her focus is on bringing the raw and living food philosophy into the mainstream. Rose's theories are described in various places in this book. In addition, her role as a leader in the raw food movement is described in the chapter on Raw Leaders.

Victoria Boutenko

Victoria Boutenko has written so many books about the raw food experience that she has her own publishing firm, called Raw Family Publishing. Victoria's books include *Green for Life, 12 Steps to Raw Foods* and *Raw Family: A True Story of Awakening*. Victoria, her husband, Igor, and their two children moved from Russia to Denver, where she accepted a position to teach political science at a college there. They were a perfectly healthy family back in Russia. Once in the United States, they became mesmerized by the American *supermarket* and vowed to try every colorful box on the shelves. In a year, Victoria gained

100 pounds and was suffering from heart arrhythmias. Her son went into a diabetic coma after eating a pillowcase full of Halloween candy and was diagnosed with Type 1 Diabetes. Her husband developed severe hyperthyroidism and heart arrhythmias.

Not able to find the answers she wanted from doctors or books, Victoria began to look to others for advice. In particular, she noticed that certain people look healthier and have a glow to them. She began asking every healthy-looking person she saw what their secret was. Shortly after she made this decision, she found the answer she was looking for. A very healthy-looking woman whom she met in the bank claimed to have cured herself of colon cancer through the raw food diet. This encounter caused Victoria to embark on what, so far, has been a lifelong journey to bring the raw and living food lifestyle to her family and to others.

Ann Wigmore

Ann Wigmore, often referred to as the mother of the raw food movement, was a big proponent of the values of wheatgrass, juicing, and sprouting. While in Germany during World War II, she first saw her grandmother treat soldiers' wounds with wheatgrass with a high degree of success. She remembered this and when Ann was diagnosed with colon cancer, she made the decision to treat herself with wheatgrass juice instead of undergoing conventional treatment. Wigmore cured herself of colon cancer using wheatgrass therapy. She eventually added other juices and sprouting (soaking seeds, nuts, and beans) as staples to her raw food diet and her natural approach to healthcare. Ann Wigmore is also discussed in the chapter on Raw Food Leaders.

Norman Walker

Norman Walker is often referred to as the father of the raw and living food and juicing movements. In particular, Dr. Walker was key in creating the first manual and then mechanical juicers. Dr. Walker tells the story in many of his books, including *Vibrant Health* and *Become Younger*, about how during his recovery from a breakdown, he was staying at a farmhouse in the European countryside. One day, while watching his hostess peel a carrot, he noticed the moistness on the inside of the peel and believed this liquid probably held the essence of the nutrients in the vegetable. He proceeded to experiment on himself by only drinking juices of all types of fruits and vegetables to determine if they would improve his life. In a short time, be began seeing positive results both physically and mentally. Relying predominately on juices for his nutrition, Dr. Walker continued to live this lifestyle until he was at least 100 years old. His age appears to be disputed and some resources record him as living as long as 115 years. Dr. Walker is also discussed in the chapter on Raw Food Leaders.

Woody Harrelson

Celebrities attract attention to the causes they promote. For that reason, Woody Harrelson's embracing of the raw food lifestyle has turned some heads. When Emmalea and I lived in New York City during her senior year in high school, we would see Woody frequently in the Candle Café, a vegan restaurant on the Upper East Side, and we suspected he was indeed a vegan. It was not until his foreword to Renee Loux Underkoffler's un-cookbook, *Living Cuisine: The Art and Spirit of Raw Foods,* that he declared himself a raw foodist. When Woody first ate in Renee's restaurant, The Raw Experience, with Gabriel Cousens, he was not a raw foodist. Although he was acutely aware of the benefits of the diet and the lifestyle, he believed he would have to sacrifice taste and substance to maintain it. The experience of the tasty and beautiful food served at The Raw Experience transformed Harrelson into a raw foodie. He says in the foreword to the book, "Anyone who has tasted Renee's pies knows that desserts can taste great and don't have to be bad for you. That is the beginning of revolutionary thinking about food."[1]

Carol Alt

In her book, *Eating in the Raw*, Alt, a former supermodel, describes her eating habits growing up on Long Island in the 1960s and 1970s. Cooked, processed foods, with a lot of pasta, white bread, and meat were staples. Alt even describes how she grew up thinking that broccoli was found in nature in bite-size chunks because the only broccoli she ever saw was from the Birds Eye frozen plastic bags that her mother dropped into boiling water. She claims she was fed but not *nourished*. Then, after becoming a model, she lost weight and found herself feeling tired, irritable, and depressed a lot of the time. She was diagnosed with hypoglycemia, and the nutritionist told her that her body was literally eating itself. Later, she developed sinus and gastrointestinal problems and a daily habit of waking up to scotch, coffee, and whipped cream, and falling to sleep with NyQuil. In 1996, she heard about a friend who was diagnosed with severe cervical cancer. Before undergoing surgery, the friend saw a doctor, Timothy Brantley, who prescribed herbs, supplements, juicing, and raw foods. After six months, she received a clean bill of health—she was cancer free. Alt herself sought consultation with the same doctor and has practiced the raw and living food lifestyle ever since.

Joel Fuhrman, MD

Dr. Fuhrman's story is about finding cure in fasting. Dr. Furhman is a staunch proponent of eating fresh, raw foods whenever possible. But he is not a strict raw foodist. He also eats cooked foods. I have included his story here because fasting is an important component of the raw food philosophy, and his is one of the best fasting stories in print. He describes the story in his book, *Fasting and Eating for Health*.

At the age of 20, Dr. Fuhrman was a world-class Olympic ice skating hopeful when he was injured. His lower-leg injury left him with chronic inflammation that could not be successfully treated by traditional medicine. When a surgeon told him that he would never walk again, Dr. Fuhrman made the decision to undergo a lengthy, medically supervised fast, as his father had done years earlier to cure his arthritis. Dr. Fuhrman fasted for 46 days at Dr. Shelton's health school in Austin, Texas. He was completely cured and went on to place third in the World Professional Figure Skating Championships. Today, Dr. Fuhrman is probably the most well-known physician managing medically supervised fasts for his patients.

Sarma Melngailis

Sarma's story is one about becoming a convert based on the appeal and taste of raw food. In her book, *Raw Food Real World*, she describes her transition. Sarma's mother was a professional chef and taught her a great deal about cooking. Eventually, Sarma graduated from the French Culinary Institute and worked in and owned a few traditional restaurants in New York City. In 2001, Sarma and a friend dined at Quintessence, a raw restaurant in New York City. They were both so impressed with the food and the large crowd this small restaurant attracted that they decided right then to start their own raw food business, now known as Pure Food and Wine restaurant, a takeaway café, and One Lucky Duck, for raw foods online. I also discuss Sarma in the chapter on Raw Food Leaders because of her contributions to mainstreaming raw foods.

Brenda Cobb

Brenda Cobb, a raw food author, gourmet, and food preparation instructor, describes her conversion to raw food in her book, *The Living Food Lifestyle*, and on her Web site www.livingfoodsinstitute.com. Brenda overcame the early stages of breast and cervical cancer without the use of drugs or surgery by following the simple principles of detoxification and nutrition. She also wiped out her allergies, acid reflux, indigestion, arthritis, obesity, liver spots, and gray hair. She adopted the principles of the Hippocrates Institute in Florida and, since 1999, has dedicated her life to teaching others about these principles. In her book she features the stories (and pictures) of many people who learned about detoxification and the raw food lifestyle from her institute. The stories are inspiring for anyone.

Returners and Extremists

In my travels through the raw diet community, I have run into individuals who describe themselves as constantly going back and forth between raw and non-raw cuisine. This is not necessarily a bad thing if you subscribe to the raw food philosophy overall and just bounce back and forth in your percentages. For example, I have heard many people describe themselves as being 75 percent raw in the cold months and close to 100 percent in the summer months. Because these individuals usually crave hot, cooked foods in the cold months, cooked foods comprise the remaining 25 percent during the winter. Raw food leaders, like Gabriel Cousens, describe this type of practice as healthful. And individuals who practice like this are likely to obtain sustained benefits from eating a diet of living foods. These folks return to a higher percentage of raw food when it makes sense for them. They stay in the healthful zone overall, but leave themselves some slack to respond to their bodies' changes.

On the other hand, extremists are individuals who go from one end of the diet spectrum to the other, eating totally raw for a month or two and then eating totally cooked and/or processed food for a few months. A good example of this was during a recent raw food preparation retreat I attended. While in class, everyone at least professed to be a raw foodist. And the only type of food available on the retreat campus was, indeed, raw. However, there were usually three or four of the students who would drive into town each evening, find a restaurant, and load up on cooked food, desserts, and sweets that they would sneak back into the retreat. You could usually hear the discussions the next morning before the instructor started class. This is not unlike other types of dietary extremes we hear about. For example, it is not uncommon for individuals to bounce back and forth from the Atkins diet to the standard American diet. One secretary to a CEO at a large corporation once told me that in her lifetime she had lost over 2,000 pounds on the Atkins diet. It took me a while before I realized that the weight she was talking about was cumulative losses over time. She was an example of a dietary extremist, also known as the *yo-yo effect.*

The problem with the dietary extremists at the raw food retreat was that they treated the raw lifestyle as a diet. As long as you believe it is a diet, you will likely react as you would to any diet: adhering for a short time (or maybe even a long time), but eventually reverting back to old habits. This practice is not a healthy one. It is possible that any health benefits gained from the raw food intake might be negated by consistently reverting back to intake of cooked and processed foods. The constant stress of change on the body can be damaging. The raw and living food lifestyle is meant to bring about ease, not stress. So practicing extreme eating is counterproductive to the lifestyle.

Determining What's Right for You

As you wade through this information, the bottom line is to decide what works best for you. Perhaps you can use these stories to determine which changes, if any, you would like to make in your own habits. In order to make any effective or sustainable change, the change must be one that is acceptable to your body. This is particularly true of any nutritional change. One of the best descriptions of why your body needs to be accepting of the changes you make is described by Roger Williams in his book, *Biochemical Individuality.* Williams first published his theory in 1956, but the book has re-emerged in popularity and was reprinted in 1998. The theory is based upon the fact that every one of us has very specific, genetically determined nutrition requirements. The biological characteristics that are unique to you may be the most important factor in the effectiveness and success of any nutritional plan you follow. Other factors to consider are your physical and emotional needs.

CHAPTER 2

Your Story

....

In the introduction and chapter 1, you read about the basic principles of the raw food philosophy as well as stories of individuals who practice the philosophy. Chapters 3 through 16 focus on raw and living food leaders and raw and living food concepts, so you can do your own analysis and make your own decisions. The best way to do any analysis is by applying a framework that incorporates your own thoughts and feelings and successful experimentation regarding what foods work best for you. You can use the information in this chapter to do just that.

This chapter provides you with a way to assess your own needs and opinions regarding food and diet. Use some of it or all of it. The questions you will answer and the structure you will create for yourself may make you more aware of your food choices. Most importantly, the structure you build will provide a dynamic, nutritionally sound, sustainable method for eating that is *unique* to you.

31

Adaptive Eating

Adaptation is the process of fitting or suiting one thing to another.[1] Adaptive eating is the process of fitting your food intake to meet your own specific nutritional needs. While some of this adaptation can be about eating foods that you enjoy, the primary focus should be on the nutritional benefits of the food to your body. Eating patterns can not and should not be adapted universally. The best combination will occur when the food you eat is good for you *and* you enjoy eating it.

Pete and Jane are both examples of maladaptive eaters. Pete does not eat breakfast or lunch on most days. When he does grab a quick lunch, it is usually an English muffin slathered with butter. He eats an enormous dinner each evening usually consisting of two to three large pieces of meat, two large helpings of mashed potatoes with butter, and a can of cooked yellow corn. Right before bed he eats a large roast beef sandwich on white bread with mayonnaise and butter. Pete loves the taste and texture of his food. As he describes his eating, "I am as happy as a pig in a poke."

Jane, on the other hand, is very vigilant about her eating habits. She begins each morning with two shots of wheatgrass, followed 15 minutes later by a green smoothie of blended kale and bananas. Lunch is spinach or another leafy green vegetable salad. And dinner is brown rice, broccoli, leafy greens, and strawberries or other fruit for dessert. Jane feels healthy, but admits to gagging almost every morning at the thought of wheatgrass. She also craves chocolate daily and may "give in" once a week, after which she scolds herself and withholds fruit the following day.

Even for someone with no nutritional training, the issues involved in both Pete and Jane's daily eating habits are apparent. Pete is happy, but is likely on a quick road to an early heart attack. Jane's food intake appears to be 100 percent healthy, but she has checked out emotionally and mentally from her eating habits. Both need to make changes. But each needs to undertake dietary and behavioral modifications that will work for them. Their changes need to occur in the following order:

1. Recognition of the Need for Change

For both Pete and Jane, unless they recognize that their eating habits are having or may have harmful consequences, they are unlikely to change. In fact, even if they recognize their unhealthy patterns, the chances of making a change are still not in their favor since habits are hard to break. At this point, it appears that Pete and Jane have developed eating patterns over a significant enough time that they are, in fact, habits.

Pete and Jane need to make changes, but the types of changes and the root cause for each is different. In Pete's case, his need to change is physical. His eating habits are either harming his body now or setting the stage for significant future problems. In Jane's case, she continues to force herself to engage in eating habits that appear to be nutritionally sound, but are emotionally harmful. The ultimate goal for each of us should be to engage in eating habits that are both physically and emotionally beneficial to our bodies and to our minds. But unless we recognize and believe that we are out of balance, we are unlikely to even consider a change.

If you want different results, you need to *change* the input. This means changing what you are eating and drinking. Any of the over 100 dietary theories that exist from macrobiotics to the South Beach diet to the Atkins diet all have a specific prescription for what will work, in general, for individuals practicing that dietary theory. The problem with most dietary theories is that they fall short of a solution when one of two things happen: either we lose our willpower to follow the specific dietary criteria or we follow the criteria and, after some time, plateau and find it difficult to maintain the results we were getting when we first started the diet.

In the case where you lose the willpower to continue to follow the diet, an overall self-assessment is probably necessary to determine whether the diet is a good solution for you. More importantly, if you have sustained the diet and reached a plateau, you can become frustrated here as well. You may not be feeling as healthy and vibrant, or you may not be losing as much weight as you were initially. Whatever result you are looking for, one thing is clear: you need to change the input to get a different result.

Even the raw food lifestyle requires periodic changes. Some describe it as periodically jump-starting the body or giving it that little nudge to get moving again. Just like our bodies, our minds cease to respond to the same exact regimen week after week, month after month. One way to avert this is to make a change, even if it is a brief one. There are a lot of interesting exercises we can do with our minds just to get moving in a different path. One of my favorites is reading several pages from a book upside down. Another is meditation. The list goes on. What is most important is that you engage in some change, even a small degree of change, to recharge your mind.

The same goes for your body. A common practice among raw foodists is to participate in weeklong juice or water fasts three to four times per year. This practice is considered both a physical and a spiritual cleansing ritual. Luckily, you do not need to take such drastic measures to jump-start your body. Some changes can create an impact in one day, depending on your current dietary practices. These examples include: a one-day juice fast, eating only fruit for a day, eating all fresh vegetables for a day (cooked or uncooked), or eliminating sugar from your diet for a week. The possibilities are endless and depend on your current nutrition, lifestyle, and goals. My yearly jump start occurs when I give up eating chocolate for 40 days during Lent. Think about what might make a difference for you, then try experimenting for a day or two at a time to see how you feel.

2. Education and Analysis

Once we determine that we need to make a change, we need information to educate ourselves. Sources of information include nutritionists, our primary care physician or other healthcare practitioner, the Internet, or one of the tens of thousands of nutrition books. It is impossible to provide a comprehensive means for determining credibility of all nutritional resources. However, the table listed below contains a few essential criteria you can start with.

3. Self-experimentation

You are your own best experiment. In fact, in the area of raw and living food nutrition, we have learned the most from individuals like Norman Walker and Ann Wigmore, who were willing to experiment on themselves to see what

Table 2.1–Criteria for Assessing Credibility of Nutritional Resources

HEALTHCARE PRACTITIONER	Does the individual have a degree or specific training in nutrition? Is she credentialed? Licensed?
BOOK TOPIC	Stick to topics on general nutrition if possible, and not about specific diets. *Integrative Nutrition* by Joshua Rosenthal is a good example of such a book. He addresses every nutritional theory in a thorough and objective way.
AUTHOR	Does the author have a degree and/or specific training in nutrition? Has the author taught and/or counseled people in nutrition in the past?
ORGANIZATIONS	If you are getting information from an organization or Web site, is it a credible resource like Mayo Clinic, Cleveland Clinic, Weight Watchers, or even WebMD?
YOUR OWN PREFERENCES	Consider your own preferences in selecting a resource. For example, if you prefer a holistic approach to healthcare, you will be more likely to listen to advice from someone like Gary Null rather than the Mayo Clinic.

worked and didn't work. The limitation with this kind of experiment is that the results can really only be applied with a high degree of confidence to the individual on whom the experiment was conducted. We can repeat it and try it out ourselves, but until we do, what was successful for Dr. Walker or Dr. Wigmore cannot be determined to be 100 percent successful for each of us.

Trying out dietary changes on yourself is beneficial in many ways. For example, if you are actively engaged in assessing your response to every change you make, you will be more aware of yourself overall—more conscious. As a result, you are more likely to be in touch with your feelings.

If and when you decide to practice self-experimentation in nutrition, there are two things to keep in mind:

(1) *Use credible resources to guide your decisions on how to experiment.* For example, you only want to change things in your diet that you know are healthy and cannot harm you. While every experiment you undertake may not produce a good result, you should not be experimenting with choices that could harm you. For example, while you may want to add wheatgrass or leafy

greens to your diet to see how your body responds, you would not want to eat the wild mushrooms growing in your backyard to see how your body reacts.

(2) *Make only one change, or undertake only one experiment at a time.* In the design of experiments, if we change more than one thing at a time, we have to design complex statistical models to try and determine the impact of one change versus the other change. In experimenting with individual nutrition, it is impossible to design any meaningful statistical analysis to weed out the significant impact. So your best bet is to make one change at a time. One of the most common nutritional missteps is taking more than one supplement, all beginning on the same day.

Let's say you suddenly have an epiphany that supplements are the way to increase your energy level and nutrition. You read several books and talk with a nutritionist. After some analysis, you decide to take cod-liver oil (for vitamin D), vitamin E, magnesium, calcium, and potassium. If you feel badly after beginning the regimen, you will need to either eliminate all supplements together and then start over one by one or eliminate the supplements one by one until, based on improved health, you are able to determine the culprit supplement. Had you initially started the process by taking one new supplement at a time, you could have determined how your body was reacting to each supplement as you went through the process.

4. Making *Small Changes*

In the United States in particular, when it comes to diet and nutrition changes, we usually want to make big changes right away. We are, after all, the country that made the acronym *BHAG* (Big Hairy Audacious Goal) popular. The most fashionable diets have been about big changes. Eliminate *all* carbohydrates. Eliminate *all* fats. Eat *only* grapefruit for a week. You get the point. The most common result of big changes undertaken too quickly is the yo-yo effect. You succeed for a short time, fail, then start over again. This is followed by more short-term success, failure, and repeating the cycle. If your objective is success, then starting with incremental adjustments and making small changes, one at a time, may be the best approach to take.

Don't get me wrong, I am not suggesting that you give up on *BHAGs*. I am, however, suggesting that you determine what all of your small steps will be to reach that goal. Your focus should be on the small steps. These steps should be easy because they come naturally to you. If you find it is difficult to stick to the small changes, then stop and go back. You need to rethink the change. It does not matter how many times you stop and go back. It could be one hundred, one thousand, or even ten thousand times. The important thing is that you keep going back and reassessing the changes until you find ones you can stick to. You are worth it.

5. Assess the Impact of the Changes

After you have made a change, determine the impact, if any, that the change has had on you. Do you feel *more* or *less* energetic? Irritable or peaceful? What other adjectives can you use to contrast the *you* before the change with the *you* after the change? The more adjectives you can select, the better. You need sufficient data about yourself to determine whether the change you made produced a desirable result. In order to assess impact you need to be aware of yourself before and after the change. You need to be patient. The impact of change can take time. Pay attention to yourself. Listen to your body, mind, and heart.

6. Keep What Works, Change What Doesn't Work

If the result produced by the change you made is what you desire, you may want to continue. Otherwise, stop the new practice immediately. And if you stop, you should be ready with your next experiment. For example, let's say you decide it's time to increase the amount of fresh vegetables you eat daily. You start by eating cucumbers every day. You notice some indigestion after about three days of eating the cucumbers, which can happen with vegetables that have seeds. In this case, you stop eating the cucumbers and replace them with broccoli. You follow this change with the same analysis and keep what works, and you continue to eliminate what does not work. Keep in mind that part of the criteria for "keeping what works" is whether you actually enjoy the food. If you do not enjoy what you eat, you will not be able to sustain the change, even if it is good for you. Sustainability is an important criteria in this structure. So if you find that, like Jane, wheatgrass juice makes you gag, eliminate it immediately from your regimen.

Biochemical Individuality

The theory of biochemical individuality was first introduced by Roger J. Williams, PhD, in his book, *Biochemical Individuality: The Key to Understanding What Shapes Your Health*. The concept is described as bio-individuality by Joshua Rosenthal, MEd, in his book *Integrative Nutrition: Feed Your Hunger for Health and Happiness*. Williams notes that the principle can be stated as, "Every individual organism has distinctive genetic background and distinctive nutritional needs which must be met for optimal well-being."[2] When Williams wrote the book in 1956, he believed our nutritional habits then were not sound. Were he still alive today, he would shudder at the content of the standard American diet.

You can take Williams' principles and use them to develop your best nutritional plan. Williams was a scientist, so he used objective, solid data to support his claims. We can use some of this data to better understand how each of us is unique and why we should approach our nutrition individually. Some of the conclusions he arrives at, using scientific data, are:

- The human stomach, liver, and colon all vary greatly in shape, position and form in different individuals (Dr. Williams provides illustrations of these differences on pages 22, 27, and 28 of his book).
- The normal range of naturally occurring substances in our blood is quite large. For example, the normal range of sodium in blood serum is 132 to 144 mg/1000 ml and the normal range of chloride in whole blood is 71 to 87 mg/1000 ml. And the range for zinc is anywhere from 488 to 1272 (page 55).
- Several studies show that there are differences in the amounts of minerals (such as calcium and potassium) needed by different individuals. Dr. Williams reviews and quotes from several of these studies in his book.

I am not suggesting that Dr. Williams' theory be taken as the most cutting-edge solution to nutrition. It is, however, a theory that appears to have a sound basis, and, it has withstood the test of time. Essentially, Dr. Williams is saying that the best nutritional practices for our own physical and mental health are specific to us. It is a good idea, and in some cases may be necessary, to partner with a physician, nutritionist, or other healthcare practitioner to help you make certain decisions. Certainly, if you want to determine whether you have any mineral or vitamin deficiencies, you will need a physician to order the appropriate tests. But you should always ask for copies of the results and ask questions about what the results mean and the steps you can take to make any necessary changes.

Variability in Nutritional Needs

Recognizing your biochemical individuality and making changes based on what makes nutritional sense *and* what makes you feel good means you are attaining optimal nutrition. Determining what makes you feel good should be relatively easy—as long as you pay attention to your body whenever you make a change. However, determining what is nutritionally sound for each of us may not be quite so easy. In order to create a sustainable plan for yourself, you need to combine sound nutritional practices with your philosophy of life. This combination will create a nutritional plan that reflects you and what you want from life. I offer two concepts as building blocks for a simple way to internalize sound nutritional practices and for developing your own nutritional road map. The first is creating your own nutritional values, vision, and mission statement (VVMS) and the second in the appendix is the federal government's new food pyramid, MyPyramid.

You Need a Map

You may not have thought a lot about nutrition or your eating habits. Some of us may think about it, but we are so overwhelmed by the numerous choices that it is hard to know where to begin. There are hundreds of diet books published every year. Nutritional diet gurus tout their philosophies on infomercials, and we hear suggestions from healthcare professionals, the government and others. The list of possibilities in nutrition and diet choices seems endless.

The obvious question to ask yourself is, "Where is the best place to start?" Start with you. Make yourself a plan and then create a map. Your plan begins with what matters most to you, your values. Your map will include these values along with your vision and mission statement (VVMS) for your nutrition. You can use your nutritional VVMS in other areas of your personal and professional life, but for now, let's focus on your nutritional planning.

Values: What Do I Value in My Food?

You need to ask yourself three key questions: (1) What qualities do I enjoy in food? (2) How do I want to look and feel? (3) Are my answers to number 1 and number 2 in harmony with each other? For example, in your response to number 1 you say that you love hot and spicy food, which gives you indigestion. Then, in your response to number 2 you say that you want to have no digestive problems. In this case, your responses to questions 1 and 2 do not harmonize with each other. So you will need to either make changes in the amount of hot and spicy food you eat or change the result you are looking for (be okay with suffering from indigestion).

Let's consider number 1, What qualities do I enjoy in food? Do you have certain textures, colors, amounts (if you are a big eater), temperatures, degrees of sweetness or spiciness, or other qualities of food that appeal to you most? Make a list of those qualities.

I like food that is (You can even mention your top two or three favorite foods here if you have some favorites.):

Keep going. If we were to only consider what *feels good on our tongue*, and our taste buds craved sweet, salty, and oily foods, we would likely be obese, irritable, and generally unhealthy—characteristics we are not likely to value. So we must include our second criteria for values, how you want to look and feel.

You are what you eat. What goes in, comes out. The clichés go on. But ultimately your food is your fuel. If you are inputting diesel fuel into a tank that requires high-octane unleaded gas, your car will not operate. It's the same with your body. Though we may not think about it much, we all know this. So how do you want to look and feel? What do you value? Some examples include high energy, lean and/or muscular body mass, a healthy glow to your skin, regular elimination, clearheadedness, and peacefulness. Make a list of the qualities you are looking for in terms of body function and appearance. For example, do you want to be able to exercise five times per week for an hour or seven days a week for 30 minutes a day? Do you want to master a particular sport or activity such as yoga, pilates, tennis, or basketball?

_____

_____

_____

Now, the third step is to determine if the values in (1) and (2) that you identified are in harmony with each other. For most of us there will be some need to reconcile the two lists. This reconciliation may be painful. You need to reconcile the two lists so you can achieve your desired outcome. If it's too frustrating, start small. You can reconcile pieces of the lists. Make small changes that you feel confident you can maintain. Remember any change *you* make is (or should be) based on *your own* values statements about what you value in how your body looks and performs. *You* are making the rules of the game, nobody else.

I will illustrate with my own brief example. The two characteristics that I look for most in food are chocolate and what I refer to as *green crunchy stuff*.

Through years of experimentation, I have found that on days that I don't eat green stuff, I feel irritable and depressed. Although not scientifically proven, I believe there must be something in the green, chlorophyll properties that impacts *my* brain chemistry. In fact, I often say, "On days when I don't eat green, I feel blue." I would be in heaven if I could exist on a diet of chocolate and broccoli. These days, it's raw, dark chocolate and raw broccoli. However, if I eat like this for more than a day, I find that my hair falls out, my fingernails break off, I am irritable, and I begin to develop other signs of protein deficiency. My values for how I want to look and feel do not include these characteristics. I want high energy, low body fat, to be alert and mentally aware, to be able to engage in at least one hour of aerobic exercise a day, shiny hair, and a healthy glow to my skin.

So, like most of us, I need to reconcile what I value in taste (what goes in) with what I value in body performance (what comes out, or how my body uses the fuel I give it). I still eat plenty of broccoli and chocolate. However, I also make sure that each day I eat raw and unsalted seeds and nuts, sea vegetables, and sprouted beans for protein. I eat a variety of colorful vegetables, to help the glow of my skin, including red peppers, carrots, yams, spinach, bok choi, and rainbow chard. I take vitamins D and E for shiny hair. And, because I do not eat meat, a main source of vitamin B_{12}, I take a vitamin B supplement.

Some days, I will eat a diet of only chocolate and broccoli. But I pay for this indulgence. My body tells me it is time to stop messing around if I want the result on my second list of values. So I stop messing around. And I go back to the list often—sometimes just to review it and reaffirm my commitment to my values statement. My list has also evolved over the decades. Ten years ago, I ate only milk chocolate (not raw or dark) and I had never heard of bok choi or rainbow chard. I ate a lot of cooked carbs like rice and potatoes, especially in the winter. Who knows, I may add these back into my diet someday, depending on how my body reacts to the food on my current list.

VVMS are meant to be dynamic documents. So I hope you will create your own three-part nutritional values listing that includes (1) what you *like* to eat, (2) what you want your body to do, and (3) a reconciliation statement.

Start with what you know about yourself. Then, as you increase awareness of your intake and your body's reactions, continue to refine each part of your values statement. Write your own analysis of whether your answers to (1) are in harmony with your answers to (2).

_____

_____

_____

Vision: How Do I See Myself?

A personal vision is a picture of yourself in the future, a picture of where you are going. One option is to adopt the World Health Organization's definition of healthy and say that you are looking for "a state of complete physical, mental, and social well-being, and the ability to lead a socially and economically productive life."[3] Your health, of which nutrition is a primary determining factor, is multifaceted. You may think in general or specific categories when creating your vision statement. Identifying categories you want to include in your vision and mission statements ensures that you are including all the relevant, important areas. Examples of categories you may want to include in your vision statement may be:

- Physical fitness level
- Quality of your hair, skin, and body composition
- Mental/emotional health
- Mindfulness/awareness
- Healthcare providers with whom you are collaborating
- Management of health information or keeping a food diary

Here is an example of a health vision statement, by Jack LaLanne, 94, who still exercises daily, "By remaining physically fit and eating healthy, I will continue teaching as many people as possible to eat healthy, whole foods and to exercise daily to lead a healthier life."[4]

Write your own vision statement for your future health and nutrition now:

_____

_____

_____

Take a Snapshot

A vision statement includes a picture in your mind of how you see yourself physically in the future. This visualization helps you achieve the vision statement that you have created. For example, the following is an example of a snapshot of future health, "To cross one last finish line as my wife and 10 children applaud." This is from Lance Armstrong, as written in his book *It's Not About the Bike: My Journey Back to Life*. He had only one child when he wrote the book.[5] Creating a snapshot in your mind and writing it down on paper can be very motivating. It can also confirm for you what you are looking to achieve. Creating a picture, a snapshot, is also a way of making a commitment to yourself. Describe the *snapshot* of yourself in the future here:

_____

_____

_____

Your vision statements may change over time just as values may change, so you will want to reassess your statement at least annually. Modify your statement to be consistent with any changes in your health, learning, or thinking.

What Is My Mission Statement?

Vision is where you are going; *mission* is how you will get there. Your mission statement weaves together your vision statement and your values. It specifically addresses how you will achieve your vision and become the person you see in your snapshot of future health. As you begin to create your mission statement, have your vision and values statements with you. Refer to the contents of your vision statement to help you sketch out your mission statement. Creating your mission statement fills in the details of how you will achieve your vision. In many ways, your mission statement is really just a merging of your values with your vision. To provide an example, here are my nutritional values, vision, and mission statement (VVMS):

Values:

(1) I like chocolate, broccoli, and greens.
(2) I value high energy, specifically the ability to participate in at least one hour of aerobic activity a day, being alert and mentally aware, shiny hair, and a healthy glow to my skin (but not from sun exposure).
(3) My reconciliation of my two nutritional values statements is that I will eat the foods I like regularly, but in moderation. In addition, I will eat organic, raw sources of protein daily and take any supplements that my body needs. Right now these are vitamins D, E, and a B vitamin complex.

Vision:

I will continue to be physically fit and healthy in a way that is sustainable. I will continue to be mentally healthy through mindfulness and awareness of my decisions and actions.

Snapshot:

I am running around the reservoir in Central Park with my grandson, who is on his high school baseball team, and my granddaughter, who is on her high school track team (I have no grandchildren yet, by the way).

Mission (Merge Values and Vision):

I will continue to be physically fit and healthy in a way that is sustainable. I will continue to be mentally healthy by being mindful and aware of my decisions and actions. I will accomplish this by eating the foods I like, but in moderation. In addition, I will eat organic, raw sources of protein daily and take any supplements that my body needs. Right now these are vitamin Bs, E, and D.

In the space below, write your complete nutritional VVMS statement:

Values:

_____

_____

_____

Vision:

_____

_____

_____

Snapshot:

_____

_____

_____

Mission (Merge Values and Vision):

_____

_____

_____

Congratulations! You have created your nutritional VVMS, a tremendous achievement and a good place to start! Hopefully you can use these basic steps to help you assess not only the information in this book concerning the raw food lifestyle, but also other diet and nutrition resources. You need to know yourself and know what works for you before you can determine whether to make a change in your diet or lifestyle. Good luck!

CHAPTER 3

Raw Leaders

· ·

As with any revolution, the raw food movement has its lead-
ers. I used three criteria to qualify an individual as a raw food leader. First,
the individual had to have contributed something unique to the raw food
philosophy. Second, the individual had to make a contribution to raw food
philosophy through writings. And, third, the individual must continue to
have an influence on the raw food movement that is sustained either by teaching, training, or pedagogy. Although Dr. Norman Walker started his raw food
diet journey in France, many claim that he was the first to bring the movement to the United States. For historical and credibility purposes, I have
also included the Essenes and European physicians who created separate but
related practices for a living food lifestyle.

The Essenes

UNIQUE IMPACT: First religious sect (second century BC) to describe, embrace, and practice the importance of fasting and eating raw foods for health.

WRITING: The Dead Sea Scrolls

TEACHING/PEDAGOGY: Today the Essene Church of Christ is lead by Nazariah (in Cottage Grove, Oregon), where the ministers follow the same traditions as the religious sect of the second century BC.

The Essenes were one of three Judaic religious groups that prospered on the shores of the Dead Sea at the time of Christ. The Dead Sea Scrolls, which most scholars believe were authored by the Essenes, contain text about the use of fasting and raw foods to attain the highest levels of mental, physical, and spiritual health.[1] Essene practitioners believe that all sickness results from disharmony. They have a holistic approach to healthcare. The Essene, or biogenic, diet is about 40 percent fresh fruit, 30 percent raw vegetables, and 30 percent grains, dried fruits, seeds, and nuts.

Dr. Max Bircher-Benner

UNIQUE IMPACT: Born in 1867 in Switzerland, Dr. Bircher-Benner was the first physician to identify the benefits of a raw diet, practice the diet, prescribe it for patients, and publish widely on it.

WRITING: *Children's Diet; Prevention of Incurable Disease; Dr. Bircher-Benner's Way to Positive Health and Vitality; Fruit Dishes and Raw Vegetables; Food Science for All and A New Sunlight Theory of Nutrition: Lectures to Teachers of Domestic Economy; Prevention of Disease by Correct Feeding*

TEACHING/PEDAGOGY: Bircher-Benner taught patients and treated them holistically, spoke internationally, and wrote until his death in 1939. Dr. Bircher-Benner's Zurich-based clinic still treats patients today.

Dr. Bircher-Benner treated first himself and then several patients with raw, pressed fruits to cure digestive disorders. He was one of the first to see and write about the connection between disease and a diet high in cooked and processed foods. He was likely the first physician proponent of disease prevention and holistic medicine.

Dr. Max Gerson

UNIQUE IMPACT: German physician who cured cancer and chronic diseases using a raw food diet in the first half of the 20th century.

WRITING: *A Cancer Therapy: Results of Fifty Cases and the Cure of Advanced Cancer; Gerson Therapy Handbook; The Gerson Primer: An Adjunct to A Cancer Therapy; A Cancer Therapy: Results of Fifty Cases;* and *The Cure of Advanced Cancer by Diet Therapy: A Summary of 30 Years of Clinical Experimentation*

TEACHING/PEDAGOGY: Dr. Gerson's book *Cancer Therapy* is still used by holistic physicians today. His daughter continues his philosophy and therapies by continuing to update them with modern medicine, write about them, and publish them.

Dr. Max Gerson was another physician who successfully used raw food practices on himself to cure debilitating migraine headaches. He later used the practice on patients with migraine headaches and found that replacing a cooked diet with raw, living foods helped almost every condition his patients had including cancer, lupus, and tuberculosis. Dr. Gerson wrote, "The secret of my treatment is that the nutritional problem is not well enough understood," and he found that sick bodies that could not handle cooked food thrived on raw, living foods.[2]

Dr. Norman W. Walker—The Juice Man (1886-1985)

UNIQUE IMPACT: Juicing

WRITING: *The Natural Way to Vibrant Health; Become Younger; Colon Health; Fresh Vegetables and Fruit Juices;* and *Water Can Undermine Your Health*

TEACHING/PEDAGOGY:	His books are still widely read, juicing is an important part of the raw food lifestyle, and the Norwalk juicer is still sold 25 years after this death.

Dr. Walker cured himself of mental and physiological problems through juicing and then used juicing as the nutritional basis of his theory of raw food lifestyle. His approach to medicine was to use a natural, nutrition-based approach. Many of his experiments, very well documented, were performed on himself. He believed that anatomy should be taught in every elementary and secondary school. He felt strongly that we should all understand our bodies, the parts, what they do, and how they interact with each other. He was probably one of the first individuals to tout the dangers of processed white flour. Chapter 14 of his book, *The Natural Way to Vibrant Health* (1972), is entitled "Destructive Food." He describes devitalized white flour products: white bread, crackers, donuts, and cakes as being highly harmful to health. He also talked about the dangers of eating cane sugar and refined sugar, but he supported eating raw honey.

Ann Wigmore—The Wheatgrass Queen (1909-1994)

UNIQUE IMPACT:	Wheatgrass juicing; sprouting
WRITING:	*The Wheatgrass Book; Be Your Own Doctor; The Hippocrates Diet and Health Program;* and *Overcoming AIDS and Other Incurable Diseases the Attunitive Way*
TEACHING/PEDAGOGY:	The Hippocrates Institute and two Ann Wigmore Institutes are still training people years after her death.

Ann Wigmore's philosophy is a simple but powerful one. Make living wheatgrass juice and living sprouts the primary part of your diet, and you will live a long and healthy life. She used the diet to cure herself of metastatic colon cancer. Dr. Wigmore and her institutes have probably treated more patients with late-stage cancer than any other holistic organization around the globe. Specific data regarding cure *rates* are not available from any of the institutes. But they continue to see patients in large numbers and anecdotal evidence seems to be quite positive. In addition to food, Wigmore also stressed the importance of sun, pure water, clean air, good posture, regular exercise, relaxation, sufficient sleep, massage, acupuncture, meditation, solitude, building a clean environment, and having pets. In her book *Be Your Own Doctor,* she describes each of these basic concepts as being essential to overall health.

Viktoras Kulvinskas—The Natural Health Pioneer

UNIQUE IMPACT: Cofounded Hippocrates Institute with Ann Wigmore
WRITING: *Life in the 21st Century; The Lover's Body;* and *Love Your Diet*
TEACHING/PEDAGOGY: Kulvinskas continues to lecture and contribute to journals.

Kulvinskas' primary claim to fame is being the key disciple of Ann Wigmore. He carried on her work at the Hippocrates Institutes, first with her, and then after her death in a fire in the 1990s. His book *Life in the 21st Century* is similar to a journal that addresses every possible topic from raw food practices in different parts of the globe. Chapters include "The Yoga of Survival in South America," "The Fruitarian Experience in Central America," and "Man-Made Dangers and the Need for Scientific Research on Raw Food."

Dr. Herbert Shelton—The Hygienic Physician and Fruitarian (1895-1985)

UNIQUE IMPACT: First medical professional to embrace raw food and fasting as a cure for disease and a necessary component for a healthy life

WRITING: *History of Natural Hygiene and Principles of Natural Hygiene; Getting Well; Gluttony is a Neurosis; Superior Nutrition; Food Combining Made Easy; Fasting Can Save Your Life; Science and Fine Art of Food and Nutrition; Food and Feeding; The Myth of Medicine;* among others.

TEACHING/PEDAGOGY: Over 100 of Dr. Shelton's books are still in print today on various topics. He headed Dr. Shelton's Health School in Texas from the mid-1900s through his death in 1987. Many current raw foodists and medical professionals studied under him, including Dr. Joel Fuhrman and Victoria Bidwell.

Dr. Shelton was a true revolutionary and considered to be on the fringe most of his professional life. His influence today is grounded in the sub-culture of hygienic physicians, who promote health through diet, exercise, fresh air, and bathing. In addition, Shelton and many of his followers eat only or mainly fruits and are considered to be *fruitarians*. Degree- and certificate-granting organizations for hygienic medicine have sprung up recently and are a common topic of the Living Nutrition magazine series. Dr. Shelton was truly ahead of his time when, in 1917, he was asked to define health and he replied, "Any definition of health that regards the individual as an isolated unit and fails to consider his function in nature must fall short of a complete definition." He further states, with a ring similar to Eastern philosophy and quantum theory that "Health is a state of wholeness . . . it does not matter how good one feels today, if his relations and conduct are not what they should be, he will not feel so good tomorrow."[3]

Rebbe Gabriel Cousens, MD, MD(H), DD–The Spiritual Leader

UNIQUE IMPACT:	Focus on nutrition and its relation to spirituality and peacefulness
WRITING:	*Conscious Eating; Spiritual Nutrition; 12 Steps to Raw Food; Rainbow Green Live Food Cuisine; There is a Cure for Diabetes: The Tree of Life 21 Day+ Program; Depression-free for Life;* and *Creating Peace by Being Peace*
TEACHING/PEDAGOGY:	He established the Tree of Life Rejuvenation Center, runs numerous workshops, and speaks around the globe.

Dr. Cousens has one of the most traditional educational backgrounds of the raw and living food movement leaders. He was a college all-star football player and then went to traditional medical school. He became interested in holistic medicine, which led him to spirituality. Dr. Cousens' philosophies extend beyond the nutritional component of the lifestyle, but they all lead back to food. For example, at his Tree of Life retreat in Patagonia, Arizona, he and his guests practice meditation and other spiritual rituals. You will find that certain foods like garlic and onions are not used in any of the raw food preparation at the retreat because these types of foods can interfere with meditation. Dr. Cousens has worked with the federal government on peace initiatives, and he has lectured around the world on his principles. Most importantly, Dr. Cousens practices the principles that he teaches at his retreat in Arizona. Dr. Cousens' most recent contribution to the raw food movement and to medical research is his book *There Is a Cure for Diabetes*. In the book, he has demonstrated the ability to reduce or eliminate diabetic patients' dependence on insulin through the use of fasting, juicing, and a very strict raw food diet that he calls a *Phase 1 Diet*. The diet is described in detail in Cousens' book. Because the diet is designed to work for individuals with diabetes; any foods with sugar or a glycemic index are not included on this diet.

Dr. Joel Fuhrman—The Fasting Expert

UNIQUE IMPACT: Bringing medically supervised fasting into the mainstream

WRITING: *Fasting and Eating for Health; Eat to Live;* and *Disease-proof Your Child: Feeding Kids Right*

TEACHING/PEDAGOGY: Fuhrman coordinates annual health getaways with the public and writes an e-newsletter.

Dr. Fuhrman is a mainstream physician who benefited personally from lengthy fasting. He now puts this same concept into practice for his patients through books and his medical practice. Dr. Fuhrman speaks all around the country and reaches many people through his Web site and his newsletter. He has a thriving medical practice, and he continues to treat patients using fasting and other natural techniques, thereby contributing to the scientific data available on these methods. Dr. Fuhrman is a strong proponent of raw foods, but he does not require a strict 100 percent compliance to the diet because he does not follow it himself.

David Wolfe—The Charismatic Entrepreneur

UNIQUE IMPACT: Standard setter for identification of raw superfoods and high-quality raw food resources

WRITING: *The Sunfood Diet Success System; Eating for Beauty; Naked Chocolate;* and *Amazing Grace*

TEACHING/PEDAGOGY: Wolfe hosts retreats throughout the year with the public and is a guest lecturer for many raw food events.

David is the genius behind the sunfood.com raw food site—probably the largest online raw food store in the world. He has consistently used his negotiation skills, obviously honed while he was in law school, to broker deals for the rarest organic raw food deals. He is said to be the only raw food supplier to have truly raw cashews, certain spices, and the best raw cacao. Regardless of the truth about these claims, he spreads the word about the value of raw food to as many people as possible. I recently heard him speak to a captivated audience of over 1,200 people for almost three hours straight. And if not stopped because of the schedule, he probably could have continued for another three hours.

Another one of Wolfe's contributions to the raw food philosophy is the creation of additional evidence that raw foods have higher energy than cooked foods. He demonstrates this in his talks and also in his books. Wolfe relies on the work of Seymour Kirlian. In the 1960s, Kirlian developed a machine that photographed the electrodynamic field, which shows the energy levels of anyone or anything that it photographs. Wolfe used these techniques to photograph cooked and raw vegetables, and he shows the striking differences between the two in his lectures and his books. The photographs of raw vegetables are much more radiant, bright, and colorful than those of their cooked counterparts. Wolfe has also photographed hands of raw and living food eaters and compared them with hands of standard American diet eaters. The raw and living food eaters' hands were much brighter than those of the SAD eaters.

Wolfe spends a lot of time in his talks and his books describing his own personal treks into the desert, the rain forest, or other natural habitats, sometimes for several weeks or months at a time. You may be apt to think the stories are all about him and his travels and accomplishments. But listen closely and you'll realize, based on his zeal to sell you on nature, that it is not about him at all. He wants his readers' or his listeners' interests to be so piqued that everyone gets the same advantages from nature that he has obtained. And, of course, a talk with David Wolfe wouldn't be the same without his mantra—*Have the best day ever!*

Victoria Boutenko—The Mom of Raw Food

UNIQUE IMPACT:	Expanding the raw food diet to the entire family
WRITING:	*Green for Life; 12 Steps to Raw Food;* and *Raw Family*
TEACHING/PEDAGOGY:	She has taught at several colleges and universities and conducts many workshops and retreats.

Victoria Boutenko is a teacher and a mother. Every mother or parent can benefit from Boutenko's own experiences, detailed in her books, about her family's plight and eventual success with nutrition and diet. She did not stop at telling the stories. She continues to teach and refine her raw food preparation methodologies and share them with anyone who is interested. Victoria and her family were the "model" immigrants from Russia who entered the United States perfectly healthy, gorged themselves on the typical SAD diet, became chronically ill, and found health in a raw and living food lifestyle. In her books you can read the details about her family members' health issues and challenges in transitioning to a raw food diet. Each member did make the transition and each was cured of his or her ailments without further medical intervention.

Brian Clement, NMD, PhD, LN—The Purist

UNIQUE IMPACT:	Bringing Hippocrates concepts into the 21st century
WRITING:	*Hippocrates Health Program: A Proven Guide to Healthful Living; Belief in All There Is; Children, the Ultimate Creation; Exercise Creating Your Persona; LifeForce; Longevity: Enjoying Long Life;* and *Spirituality in Healing and Life*
TEACHING/PEDAGOGY:	Clement teaches at the Hippocrates Health Institute weekly throughout the year.

Clement has not strayed from the pure living food diet. He has focused on including living foods in his diet and has marginalized foods like raw sweeteners, nuts, and seeds. Hippocrates is often the last-resort solution for cancer patients who have been told they are terminal and patients with other chronic disorders. And although statistics have never been published, anecdotal information from both patients and doctors alike tout the wonders of the Institute. Like any good scientist, Clement knows that if you change one ingredient in the mixture, you may negatively impact the outcome. And if your outcome is getting good positive results as the Hippocrates Institute seems to have been getting for the past 50 years, why mess with a good thing?

Sarma Melngailis

UNIQUE IMPACT:	Bringing raw food to mainstream America through fine dining
WRITING:	*Raw Food Real World*
TEACHING/PEDAGOGY:	Sarma writes the OneLuckyDuck.com blog and newsletters.

My husband, Joe, and I have taken many friends and relatives to Sarma's Pure Food and Wine Restaurant. They are consistently pleasantly surprised by the atmosphere. They are also caught off guard by the amazing food—its presentation and its taste. We have had meat-eating friends feast on one of Sarma's Portobello mushroom entrées and swear it was as good as a steak dinner. Although the restaurant serves only raw vegan food, Sarma is the first to admit that most people who frequent the restaurant are not raw foodists or vegans. It is rumored that most of the CEOs of Fortune 500 companies have dined at Pure Food and Wine. Sarma's contributions in moving raw and living food into the mainstream are greater than they may appear. Minimally, tens of thousands of people who may never have known what the phrase *raw food* meant before visiting Sarma's place can now spread the word with positive accolades. And what's even better—she shares it by publishing all of her recipes in her books or on her Web site.

Matthew and Terces Engelhart

UNIQUE IMPACT: The first successful chain of raw food cafés: *Café Gratitude*
WRITING: *I Am Grateful Cookbook*
TEACHING/PEDAGOGY: The Engelharts host seminars, retreats, and cooking classes.

Matthew and Terces Engelhart have done for raw restaurants what David Wolfe has done for raw food supply with a lot of love and, of course, gratitude. The food is creative and tasty, but not too overstated. And in true raw food philosophy style, the Engelharts share their recipes for all of us to at least try to reproduce. They are willing to share the secret to their ability to reproduce raw food successfully enough to open café after café. Their actions should make anyone even remotely interested in the living food lifestyle grateful for the Engelharts and their contributions.

David Klein—The Publisher

UNIQUE IMPACT: First successful journal dedicated to raw foods
WRITING: *Self-healing Crohn's and Colitis; Self Healing Power! How to Tap Into the Great Power Within You; The Fruits of Healing; The Seven Essentials for Overcoming Illness and Creating Everlasting Wellness;* and *Your Natural Diet: Alive Raw Foods*
TEACHING/PEDAGOGY: Klein teaches workshops and seminars and continues to publish and contribute to the *Living Nutrition* magazine twice per year.

Dr. Klein's decision to publish a legitimate raw and living food magazine was gutsy indeed. Published two or three times a year, the magazine boasts regular columns like *Ask the Nutritionists* and *Apples & Oranges, Peas & Carrots*, and *Raw Recipes* by notable contributors. Leaders like Gabriel Cousens and Dr. T. Colin Campbell have been interviewed for the journal. The journal adheres to a strict philosophy that embraces the importance of fruit over vegetables and de-emphasizes sweeteners. The magazine has a strong *living* foods slant to it. It is a great resource for ongoing information about the raw and living food community.

Cherie Soria—Trainer of Chefs of the Future

UNIQUE IMPACT: Directs the only organization dedicated to training raw food chefs

WRITING: *Angel Foods: Healthy Recipes for Heavenly Bodies;* and *The Raw Food Diet Revolution: Feast, Lose Weight, Gain Energy, Feel Younger!*

TEACHING/PEDAGOGY: Soria established Living Light Culinary Arts Institute.

Cherie has trained many of the top raw and living food chefs, which is why people often refer to her as the mother of raw gourmet cuisine. Cherie studied with Dr. Ann Wigmore and learned the principles of using whole, live foods to aid in healing and rejuvenation. Inspired and motivated, she went on to develop gourmet raw culinary arts by combining the two contrasting worlds of gourmet vegetarian cuisine and raw vegan simplicity that is known today internationally as gourmet raw vegan cuisine. Cherie says, "I experience great joy connecting personally with students and helping them to become the best that they can be—not only as raw food chefs, but as radiant, inspiring teachers. The key is to accept people where they are on their path and to help fuel the light that shines within them!"

Bob Dagger—The Radio Man

UNIQUE IMPACT: First raw food grocery store; first raw food weekly radio show

WRITING: *High Vibe* newsletter

TEACHING/PEDAGOGY: Dagger has weekly radio shows, weekly classes at the High Vibe store in New York City, and individual coaching sessions; he also does blogging. Bob writes for many magazines in Japan and Europe.

In the early 1990s, in New York City's East Village, Bob Dagger opened the city's first all-raw vegan food and supply store. The store is still going strong today and now includes an online business. He does radio shows and podcasts on a weekly basis, during which he addresses many different raw food lifestyle topics from rebounding to cleansing. He was a pioneer in the raw food supply business.

Natalia Rose—The Raw Food Realist

UNIQUE IMPACT:	Bringing the raw food lifestyle into mainstream America
WRITING:	*The Raw Food Detox Diet; Raw Food Life Force Energy;* and *The New Energy Body*
TEACHING/PEDAGOGY:	Rose teaches, coaches, and blogs.

Rose is a raw food *realist* because she decided, based on her own experience and her interviews with many nutrition clients, that the typical American would not embrace a pure raw food diet. So Rose has worked a list of foods into her recommended transition diets for people who want to become familiar with the raw food diet in a way that will allow for sustainability. Foods that Rose includes in her raw diet, which she details in the *Raw Food Detox Diet*, include maple syrup, eggs, whole grain bread products, raw goat cheese, chocolate (not raw), fish, organic meats, and organic butter and cream. Keep in mind that these are not foods that you will find on the typical raw food diet. But Rose's focus is on making the raw transition possible for as many Americans as possible.

David Jubb

UNIQUE IMPACT:	Brought the science of nutrition to raw food recipes
WRITING:	*Jubb's Cell Rejuvenation, LifeFood Recipe Book, and Secrets of an Alkaline Body*
TEACHING/PEDAGOGY:	He continues to manage his raw food café, *Jubb's Longevity,* and cooking classes, along with his wife, Annie Jubb, in their East Village location in New York City.

David Jubb is a PhD neurophysiologist from NYU. He has fused biology and physics with the living food lifestyle. Some of the theories that he pioneered or refined are used throughout the living foods community. These include the theories of cell rejuvenation, the impact of crystalline water structure, and vibration. Dr. Jubb's protocols for fasting and cleansing are quite rigorous, but proven. His book *Cell Rejuvenation* is not an easy read, but it does show that he practices what he preaches. Ultimately, Dr. Jubb has brought credibility to the raw food movement through his scientific approach.

CHAPTER 4

Raw Food Theory

··

A theory is a system of ideas or statements explaining something.[1] Raw and living food theory claims that uncooked food produces greater health benefits than cooked food. Based on its history, leaders, and the current documented practices, the following six concepts appear to be the principles of the raw food or living nutrition theory. (1) Eat food that is alive with its enzymes still intact. (2) Eat food that is fresh and organic. (3) Eat vegan food, no animal products. (4) You reflect the vibration, or energy level, of what you eat. (5) Juice some percentage of your fruits and vegetables regularly. (6) Follow food-combining rules. Details of each concept are addressed in detail later in the chapter.

A theory is meant to be tested to prove or disprove its validity. Many of the individual leaders discussed in chapter 3 engaged in testing raw food theories on themselves with success. Physicians Bircher-Benner and Gerson treated their patients using raw food principles with much documented success. A theory needs to be tested using rigorous scientific testing methods with positive, consistent outcomes time after time. Although there is abundant and reliable case study data available, clinical trials and other similar scientific studies have not yet been conducted to test the raw food theory. The good news (for supporters of raw food theory) is that there does not appear to be any solid evidence *disproving* the benefits of the theory.

While we wait for studies to be conducted at some point in the future, we can read, analyze, and possibly even try out some of the principles of the raw food theory in our own lives. I overheard a conversation recently about raw food. A man was describing the raw food diet to his companion. His explanation contained many of the six principles listed below. His friend responded, "It sounds like just another diet fad to me." The man (who was not a raw foodist) responded, "Maybe, but when you think about it, how can something that is fresh, organic, and good for you be a fad?"

The Six Principles of the Raw Food Dietary Theory

1. Eat Food that Is Alive with Its Enzymes Still Intact

Raw or living foods are believed to be the best for your body because they still contain their natural enzymes, which are destroyed in the cooking process. Food with intact enzymes are hypothesized to be much less of a burden on the digestive system than foods without enzymes. When we eat cooked foods it requires our body to go into significant enzyme-production mode, using a lot more energy to digest. Food that is alive can be one of two varieties under this theory. First, like fresh fruits and vegetables, it can be raw or uncooked. Second, food can be dehydrated up to a temperature of *about* 112° F. I use the word *about* here because there is some disagreement in the community as to the exact maximum temperature at which living food enzymes remain intact. Some say temperature cannot be higher than 110° F while others claim 120 is the maximum. Clearly here is an area ripe for scientific testing.

Let's look first at the claims about the values of enzymes and then explore the dehydration process more. Most of the published information on enzyme benefits to the human digestive system in the form of a raw food diet come from Dr. Ann Wigmore (*The Hippocrates Diet and Health Program*), Dr. Edward Howell *(Enzyme Nutrition: Unlocking the Secrets of Eating Right for Health, Vitality and Longevity)*, and Dr. Gabriel Cousens *(Conscious Eating)*. In fact, in their books, most of these authors reference the same studies and in some cases even reference each other. There are three concepts that reappear in the literature about enzymes in food and their impact on nutrition.

First is the necessity for enzymes in breaking down food. Vitamins were first identified as important nutrients in the mid-1930s. Some time after that, the importance of minerals was identified. Today we continue to be bombarded by the nutrition megaindustry with a multitude of vitamin and mineral supplements, pills, powders, drinks, and even individual elements like vitamin C or riboflavin added to our foods. Enzymes and their importance took a little longer to catch on in the nutrition megaindustry. In his book *Enzyme Nutrition,* Dr. Howell provides a wonderful analogy for our body's nutritional cycle. He states, "In order to build a product a factory needs materials of various kinds, such as steel, brass, plastics, and so on. But these would not be able to realize final form without the worker. And foremen are also needed to direct the workers. In the living body, protein, fat, carbohydrate, vitamins, and minerals are the materials to work with. The enzymes are the workers, and the hormones are foremen."[2]

Second is the presence of enzymes in foods that are not cooked that aid the body in its digestion. In *Conscious Eating,* Gabriel Cousens states that eating raw foods is the number-one activity that preserves enzymes and maximizes health.[3] He also talks about the fact that humans are the only living things that eat cooked foods. Further, he provides evidence that when animals are fed cooked foods, they also begin to suffer from chronic degenerative diseases, similar to those common to humans.

Dr. Howell describes a process called *predigestion,* or how enzymes in raw foods actually digest their own ingredients. For example, a banana is about 20 percent starch when it is green. The enzyme amylase converts the banana into 20 percent sugar when the fruit is in warm temperatures. About a fourth of this sugar is glucose, and it does not require further digestion by the human body. The natural conclusion is that the more raw foods we eat that contain their own living enzymes capable of predigestion, the less strain on us, and this gives our digestive systems a break. This inactive time allows the body to tend to important things like staying healthy, instead of trying to figure out how to take apart and detoxify itself of cooked and processed foods.

Third is the presence of enzyme inhibitors in certain foods that can actually interfere with your body's own enzymes. According to Dr. Edward Howell and others, enzyme inhibitors are present in seeds and nuts grown in trees like pecans, walnuts, Brazil nuts, and filberts. Raw seeds and tree-based nuts do not bring the positive predigestion quality (described above) with them, and they can actually *interfere* with our body's own enzyme production. Herein lies possibly one of the biggest *warnings* about problems with a raw food diet. However, there is a solution. Both Howell and Wigmore talk about the fact that enzyme inhibitors in seeds and nuts can be changed from harmful inhibitors to helpful enzymes by germinating them, more commonly known as sprouting. Sprouting involves soaking seeds, nuts, or beans for several hours until the food begins to develop a little *tail-like* protrusion on one end. The sprouting of seeds neutralizes or inactivates the enzyme inhibitors. The *good* enzyme activity is at its height when the *tail* or the *sprout* is approximately ¼ of an inch long.[4] Again, most raw food experts recommend that only 10 to 20 percent of your intake should come from raw seeds and nuts.

Dehydrating foods at a maximum temperature of somewhere between 100° F and 120° F is the last part of the principle of eating foods that are not cooked, with enzymes still intact. Unless your oven will allow you to program it to this low level (most don't), you will need to purchase a dehydrator for the purpose of dehydrating raw foods. Why would you want to do this? There are many raw foodists who do not use dehydrators at all. They exist on raw fruits and vegetables and a very small proportion of nuts and seeds. But, most of us would become quickly bored with such a diet. So the primary reason to dehydrate food is to add variety and interest to raw foods. This helps to avoid boredom and increase compliance to the lifestyle. A second reason for dehydrating raw foods is to increase the life span of the food. Once most raw foods are prepared, they should be eaten within a day or two to ensure you get the highest nutrient value. However, dehydrated foods, depending on what they are and who you ask, can last for weeks or months. For example, both dehydrated fruit and raw dehydrated crackers made from seeds can be stored for several weeks and retain their freshness.

You can find details on how to dehydrate foods in many of the well-known raw food preparation books that are listed in the appendix. However, I would like to share a few anecdotes with you about the dehydration process to help you get a sense for how this all works. First, depending on what you are dehydrating, the process can take up to 36 hours! This is particularly true when you are dehydrating nuts and seeds to make bread or breadlike foods. The minimum amount of time for dehydrating almost any food is usually five or six hours. So you need to plan accordingly for this activity.

Second, there is no *ideal* amount of time for most dehydrated foods. This is a particularly difficult concept to get used to when you have previously existed in a world of setting your microwave for two minutes and 20 seconds according to the explicit directions on the package. Most raw food preparation directions will state something like "Dehydrate at 110° F for about six hours, flip over and dehydrate on the other side for about another six hours." I was quite irritated by this process for about a year, until I visited Gabriel Cousens' Tree of Life Retreat for raw food preparation courses. The instructors were clear from the start that it is important to *be one* with your food (a concept new to me), to taste it during the prep process, and most importantly, not to *freak out* if you make a mistake or are out of an ingredient. There, they taught us to be creative using certain rules of food-combining principles (described in number 6 below). But most importantly, we needed to feel good about our food preparation because the *vibration* we embodied during food preparation would be passed into the food. This concept of vibration is discussed in number 4 below. Since visiting the Tree of Life, I check my dehydrated foods more frequently and know that it is not a problem if I pull the food out a little early or leave it in the dehydrator a bit longer.

Third, food that contains too much moisture can ferment during the dehydration process. Fermentation produces *friendly* bacteria that your body needs to make your digestive system more efficient and healthy. However, you may not like the taste. Sometimes, cooking with too much moisture in the food at too low of a temperature can cause unfriendly bacteria and even mold to form. The Tree of Life teachers addressed this during our classes by recommending that, during the first hour of dehydrating, you turn the temperature up to about 140° F. In the world of raw food, this seemed like an ultrarevolutionary idea. *140° F?* Wouldn't we be killing the enzymes? The response I received was that no, the focus would be on absorbing the liquid, not overcooking the food. I recommend that you refer to the appendices for more resources on raw food preparation and dehydration if you have an interest in exploring this further.

2. Eat Fresh and Organic

When you read this title, you may have thought, "Of course!" But in our hurried society, it is not only difficult to remember and practice these principles, it may be difficult to *find* foods that fit the bill. Let's talk about organic first, and then about fresh. There are some grocery stores, like Whole Foods, who pride themselves on delivering organic produce and other *whole foods* to their customers. Even in a store like Whole Foods, you need to read the signs because not *all* produce is organic. All produce is labeled, or should be clearly labeled in every store as *conventional* or *organic*. In the United States, the National Organic Program (NOP) is the federal agency governing organic food. In October 2002, the Organic Food Law was passed and is administered by the Department of Agriculture (USDA). Organic food is produced by farmers who emphasize the use of renewable resources and the conservation of soil and water to enhance environmental quality for future generations. Organic meat, poultry, eggs, and dairy products come from animals that are given no antibiotics or growth hormones. Organic food is produced without using most conventional pesticides, without using fertilizers made with synthetic ingredients or sewage sludge, and without bioengineering or ionizing radiation.

Before a product can be labeled *organic*, a government-approved certifier inspects the farm where the food is grown to make sure the farmer is following all the rules necessary to meet USDA organic standards. Companies that handle or process organic food before it gets to your local supermarket or restaurant must be certified, too.

Products labeled *100 percent organic* must contain only organically produced ingredients. Products labeled *organic* must consist of at least *95 percent* organically produced ingredients. Processed products that contain at least *70 percent* organic ingredients can use the phrase *made with organic ingredients* and list up to three of the organic ingredients or food groups on the principle display panel.

Let's talk about the term *sustainable* for a moment. We have started hearing this term used to describe food and wine in places like New York City, where certain types of organic produce are difficult to obtain, especially during the winter months. The term *sustainable* as it applies to food was defined in 1990 by the Food, Agriculture, Conservation, and Trade Act (FACTA). The term *sustainable agriculture* means an integrated system of plant and animal production practices that have a site-specific application and that will, over the long term: (1) satisfy human food and fiber needs; (2) enhance environmental quality and the natural-resource base upon which the agricultural economy depends; (3) make the most efficient use of nonrenewable resources and on-farm resources and integrate, where appropriate, natural biological cycles and controls; (4) sustain the economic viability of farm operations; and (5) enhance the quality of life for farmers and society as a whole. So, sustainable is a nice, warm and fuzzy concept. However, the standards seem fairly fluid and the possibility that anyone with *good intentions* regarding their food production could be labeled a member of the *sustainable* society may not be too far-fetched. What is most important is that you have an understanding of the terms *organic*, *sustainable*, and *conventional*, and you are able to identify and analyze your food purchases accordingly.

Now let's talk about *fresh* and what the term means to us and for us. Fresh can be defined as "retaining its original qualities; not deteriorated or changed by lapse of time."[6] Under this definition, the freshest food would be food that you eat as soon as it is pulled from the ground, tree, or bush where it is growing. So one might imagine a personal garden or orchard to be among the most prized possessions of someone in search of the freshest-possible food. This may not be an option for all of us, so we settle for the next-best thing. For some of us, this may be buying organic at the local Whole Foods store. However, there are many farmers who provide fresh, locally grown produce via farmers' markets or even delivery services to individuals in their communities. You can find these farmers' organizations through the Internet or local advertisements. I use an organization called Door to Door Organics (www.doortodoororganics.com). They are a national organization that connects with farmers in various communities to deliver fresh, locally grown produce to area residents. Each week of the year, I receive two large boxes of fresh, locally grown fruits and vegetables delivered to my door. I couldn't feel better about the money spent or the food as I am preparing it, sharing it, or eating it. Some questions you might want to consider are: How fresh is my food? How much of my food is fresh? If not very much, do I care or want to change this? How can I find fresher food in my community?

3. Eat Vegan Foods, Free of Animal Products

It's probably important to acknowledge that a raw and living food diet can encompass different types of foods. The diet I describe in this book that fits into the raw and living food lifestyle embraces the principles of veganism. However, there are raw food diets that include many animal products. For example, you can read about eating raw meat or raw fish in Carol Alt's book, *Eating in the Raw.* The most well-known version of the raw food diet that is not vegan is the one espoused by the Weston A. Price Foundation. Dr. Price was a dentist in the early 1900s who studied the teeth, health, and eating habits of isolated but very healthy populations like Eskimos, Irish, Swiss, and Africans. While the diets of these people varied significantly, he found they did contain several commonalities. Among them were: (1) they ate locally, (2) they ate liberal amounts of seafood or other animal proteins and fats, (3) they ate fats, meats, fruits, vegetables, legumes, nuts, seeds, and whole grains in their whole, unrefined state, and (4) all primitive diets contained some raw food of both animal and vegetable origin. One consideration in analyzing these findings is the purity of the animal products in these isolated areas during the first part of the 20th century. Today the Weston A. Price Foundation supports raw diets that include animal proteins and fats as well as fresh, raw vegetables, fruits, grains, and sprouts. They also are big proponents of raw dairy products, including meat and cheese. Enzyme nutrition in dairy products is destroyed in the pasteurization process, so they recommend finding local farmers who can and will sell raw dairy products. You can read more about the studies, the philosophy, and the foundation in Sally Fallon's book *Nourishing Traditions: The Cookbook that Challenges Politically Correct Nutrition and the Diet Dictocrats.*[7]

Veganism is defined by the American Vegan Society using six principles that spell out the word *ahimsa,* which is the philosophy of nonviolence practiced by Mahatma Gandhi. The six principles are: (1) abstinence from animal products, (2) harmlessness with reverence for life, (3) integrity of thought, word, and deed, (4) mastery over oneself, (5) service to humanity, nature, and creation, and (6) advancement of understanding and truth.[8] Of course, not all of these principles can be applied directly to diet and nutrition, but you can get a basic idea of what being a vegan means. The number-one component is abstinence from all animal products. This includes all dairy because it comes from cows, goats, or sheep. This is often the difference between individuals who refer to themselves as vegetarians (many of whom eat dairy and/or eggs) and vegans. Vegans do not eat honey because it is created by bees using their own saliva. The issue of honey, however, is a thorn in the side of the raw food community because raw honey provides so many nutrients and enzymes. This is discussed in more detail in the chapter on healthy conflict. Like the raw and living food lifestyle, veganism goes beyond just food. To be a practicing vegan, you don't wear leather goods or use any toiletries or other products (like soap) made from animals. We discuss veganism in more detail in the chapter on ethics.

4. You Reflect the Vibration of What You Eat (Thanks to Quantum Physics)

You may wonder what vibration has to do with raw food, or for that matter any food. And possibly what it all has to do with quantum physics. Well, for starters, each of us vibrates about 570 trillion times per second.[9] And, in fact, everything has a vibrational quality to it. This vibration, or wavelike quality, is at the heart of quantum theory, which suggests that all vibrations, at some level, can have an impact on all other vibrations. Let's turn to the relevance of vibration to raw food.

I have heard raw foodists commonly talking about the importance of the *love* or the *intention* that goes into food preparation. In some raw food cafés or on some packaged raw food labels, you may find one of the ingredients listed is *love*. When I first heard it, the concept just seemed a little flower child-like to me, and I accepted it as such. However, as I began to learn from raw food experts, I repeatedly heard the reference to the *vibrational* quality of food. The first to mention it was Natalia Rose. She dedicates an entire chapter to the vibrational quality of food in her *Raw Food Life Force Energy* book. Rose says that everything vibrates and, more importantly, things that are alive have more rapid vibration rates than things with little or no life. Her point is that if we fill our living bodies with low-vibration foods, we will slow down our own vibration and distort our energy levels.[10] One of the most striking statements Rose makes is that animal flesh "carries with it the vibration of death and fear so it does not count as a harmoniously vibrating food (even if it is raw)."[11]

In the 1960s, Seymour and Valentina Kirlian developed a machine that photographed energy patterns. They used the machine to make pictures of the energy pattern of two leaves. One of the leaves had a disturbed and uneven pattern. This leaf had come from a diseased plant that was soon to die. Seymour took a picture of his own hand and found his energy field also seemed to be disturbed and uneven. Soon after, he became ill. The Kirlians then made photographs of persons in poor mental and/or physical health and discovered that these photos always reflected disturbed and uneven patterns in their energy fields.[12] In his book *Eating for Beauty*, raw and living food entrepreneur David Wolfe includes several Kirlian photographs of vegetables in their cooked and uncooked states. Consistently, the photos of uncooked vegetables are brighter, more vibrant, and have significantly more light rays (representing energy levels) emanating from the vegetable than the cooked vegetables.

The concept of everything (whether living or not) vibrating, or having a measurement of energy, was birthed by quantum physicists. Wavelike patterns have been identified in all matter through breakthroughs made by 20th century physicists like Max VonLue, Ernest Rutherford, Niels Bohr, and Max Planck, among others. As Fritof Capra describes in *The Tao of Physics,* at the subatomic level all matter (living or nonliving) dissolves into wavelike patterns that represent probabilities of interconnections. More specifically, subatomic particles have no meaning as isolated entities, but can only be understood as interconnections between the observer and the matter being observed. Capra states, "Quantum theory thus reveals a basic oneness of the universe."[13] The physicists who contributed to the film *What the Bleep Do We Know!?* describe this phenomenon as "All particles are intimately linked on some level that is beyond time and space." Quantum physics "paints a picture of the universe as a unified whole, whose parts are interconnected and influence each other."[14]

When you consider basic nutritional theory through the statements and findings of raw food proponents like Natalia Rose and David Wolfe, in light of Kirlian photography and more importantly, quantum theory, you can reach the following conclusions without too much of a stretch. (1) We (and all matter) give off a certain amount of energy in the form of vibrations (2) Living things give off higher energy than nonliving things (3) If we consume living things, we are taking more energy into our bodies than if we consume nonliving things (4) If we eat higher energy food, we are more likely to have higher energy than if we eat low or no-energy food. What you do with this information is up to you. It is worth your time to assess whether your daily food intake has high, low, or no energy and consider the implications.

5. Juice Some Percentage of Your Fruits and Vegetables Regularly

Norman Walker and Ann Wigmore were the first to tout the advantages of juicing your food. One of the greatest advantages of juicing is that the juicer, by separating the nutrient-filled liquid from the fiber in the vegetables and fruits, does a lot of the digestive work for you. As a result, you give your digestive system a break while concurrently bombarding it with all things that are good for it (and you). The downside to juicing is the expense of a juicer, the time it takes to juice, and if you are not creative in your combinations or adding spices, the juice may be unbearable to drink. You can spice up the taste of green juice by adding lemons, ginger, cilantro, or even parsley.

Norman Walker stated that the two reasons for drinking fresh vegetable juices (nothing from a can or bottle) were (1) to obtain the highest quality organic water from the vegetables, and (2) to extract all of the minerals and vitamins from the plants so your body can easily process them.[15] Ann Wigmore was a big proponent of the benefits of wheatgrass juice. The human body cannot process wheatgrass in its totally raw state. Wheatgrass contains 20 of the 21 amino acids used as the building blocks of protein. Whether the body can utilize these in an efficient manner has not been scientifically proven.

Ann Wigmore claims that wheatgrass contains vitamins A, B, C, E, and magnesium, potassium, and calcium. Wigmore, through her own research on herself, patients, and thousands of guests at her Hippocrates Institute was able to use wheatgrass to improve or eradicate many types of diseases. You can read about her experiences in *The Wheatgrass Book*. Gabriel Cousens talks about the benefits of juicing in *Conscious Eating*. He points out that juices bring an alkaline force into the body that helps to neutralize the toxic acidity from which most people suffer. Alkalinity and acidity represent pH levels of the body and are discussed in detail in the chapter on pH. This speeds the recovery from disease by supporting the body's own healing activity and cell regeneration.[16]

A possible alternative to juicing is drinking green smoothies. Green smoothies, first made popular by Victoria Boutenko in her book *Green for Life*, involve the blending (as opposed to juicing) of greens usually with bananas or some other type of fruit. The use of fruit makes the drink palatable. And including greens provides both vitamins and minerals as well as fiber because you are not juicing out the fiber in the blender. The benefit you get from juicing that you don't get from a smoothie is the separation of the high-nutrient liquid from the fiber that assists in your digestive process.

Juice fasting (as well as water fasting, discussed in the chapters on water and healthcare) is a common practice among raw foodists. The value of juicing to the body appears to be fairly clear based on years of personal experimentation and documentation by physicians and scientists like Norman Walker, Gabriel Cousens, and Ann Wigmore as well as Joel Fuhrman and David Jubb, not to mention non-raw foodists like Jack LaLanne. Juice fasting is usually undertaken for both physical and spiritual cleansing. Dr. Cousens states that most people, with the exception of those 10 pounds or more underweight or with severe wasting disease, can benefit from fasting. Fasting, he states, has been known to alleviate many diseases and is actually quite safe.[17]

6. Follow Food-Combining Rules

The basic principle of food combining claims that the combination of different foods eaten at one meal can either help or hinder digestion and overall health. William Howard Hay introduced food combining to the United States in 1911. His approach was based upon the ideas at the time regarding the degree of alkalinity required to digest the food in the stomach and the pH of food itself. Herbert M. Shelton also contributed a food classification based on the type of nutrients in products. Shelton categorized foods into three groups—protein products, carbohydrate products, and "neutral" products—and recommended eating proteins and carbohydrates at separate meals.

Today, you can obtain suggestions regarding food combining from many of the raw food experts referenced in this book. Natalia Rose addresses food combining in her books *Raw Food Detox Diet* and *Raw Food Life Force Energy* as the quick-exit principle. She states that "the foods and combinations of foods you eat must make a quick exit in order for them to contribute to your Life Force Energy quotient."[18] Others, such as Brian Clement from the Hippocrates Health Institute have developed very detailed and strict guidelines for food combining that they pass on to their patients and guests at the Institute. Gabriel Cousens suggests that the simplest rule of food combining is "to eat a food, or combinations of foods, that in our direct experience are easiest for us to digest and thereby maintain our life energy and enzyme reserve."[19] The bottom line is that if you eat foods according to the rules of food combining, they will leave your body in a shorter period of time than if you didn't follow the rules. And the rules can come from a group of nutritionists or from your own body. It's just important to get the food out of your body as quickly as possible, three to four hours maximum in the stomach.

For example, fresh fruit on an empty stomach leaves the stomach in 15 to 20 minutes, while vegetables take about an hour. It takes about three to four hours for each of the following foods (alone) to exit your stomach: a serving of pasta, whole grain cereal, fish, or chicken breast. Chicken with broccoli would still only take about three to four hours to exit the stomach. However, chicken and bread or chicken and fries takes about eight hours to exit the stomach.[20] The more combinations you eat simultaneously, the longer it takes the food to exit from your stomach.

Gabriel Cousens suggests that you note what combinations of foods you eat and then, one to three hours later, determine if you have any of the following symptoms: gas, bloating, headache, feeling full, feeling hungry, indigestion, food feeling like a lump. If you have any of these symptoms, you either need to improve the quality of your food combining and/or decrease the quantity of food. The following are basic principles of food combining. All are driven by the basic premise that you should not eat foods simultaneously that will cause the total digestion time of the food with the longest digestion time to be increased. These rules are as follows. (1) Do not eat carbohydrates and proteins together. Proteins combine with vegetables and carbohydrates combine with vegetables, but don't mix the two. (2) Fresh fruit should only be eaten alone on an empty stomach. Fruits eaten with other foods, especially vegetables, can cause a backup in the digestive system and produce bloating. (3) Combine nuts with only other nuts, seeds, dried fruits, bananas, and all raw vegetables. (4) Dried fruits combine only with other dried fruits, avocados, bananas, nuts, and all raw vegetables.[21] Certain foods like avocados, lemons, and papaya seem to go well with any type of food.

As you can imagine, the *rules* of food combining can become rather cumbersome and complicated. To avoid frustration, if you do have an interest, the best place to start with basic combining concepts is probably Rose's book *Raw Food Life Force Energy*. As with all of the concepts in this book, I am presenting information in an objective manner, making it possible for you to analyze and decide what, if anything, you want to do with it. There may be one topic, like eating fresh and organic or figuring out how to get more enzymes into your own diet that may interest you. You can begin experimenting using the information in this book and then referencing the resources provided in the appendices. As you move along, remember to refer back often to your Nutritional VVMS statement as a guide for your own actions. It can keep you focused, keep you on track, and help you make some positive changes in your nutritional habits.

THE *food*
·················

CHAPTER 5

The Great Chocolate Debate

The eating of fine chocolate is a tradition passed down in my family. I learned it from my father and his gourmet hot-fudge sundaes and the boxes of Russell Stover candies he would buy me for Valentine's Day, my birthday, and other special occasions. I have taste-tested almost every type of chocolate from every country over the past 40 years—Belgium is my favorite by the way. By the time my daughter was four, I was already passing on the tradition to her when we created a game out of *appreciating* the Godiva chocolate candy bars we had purchased for ourselves one day. The main rule to the game was that you were not allowed to bite or chew the chocolate, you had to let it melt in your mouth. Luckily for us, Godiva was still using cocoa butter back then and not the oily substitutes that many chocolate bars contain today to enhance the shelf life. Since cocoa butter melts at about the same temperature as the body, the chocolate bars, which tasted divine, melted rather quickly. Today, my chocolate tastes have turned to raw chocolate.

Raw chocolate, made from raw cacao beans, is always dark and at least 75 percent cacao (pronounced ka-cow), not to be confused with cocoa, the sweetened, cooked form of the cacao bean. My favorite recipe is an organic, raw, vegan super chocolate fudge bar made with raw cacao, raw cacao butter, raw coconut oil, raw agave nectar, organic raw whole vanilla bean, organic raw cayenne, and organic Himalayan crystal salt. The bars are made by Jon-Michael Kerestes, owner of Love Street Living Foods. When I interviewed Jon-Michael, he said that he became a raw chocolatier after he saw people's reactions to the food. He brought some raw chocolate bars to a party in Los Angeles and saw that people couldn't stop at just one piece, and eventually they were competing for the last few pieces. "If raw cacao can create this reaction in people *and* it's good for you," he thought, "this is a business I must start." And so he did. And many of us are grateful for that.

The great chocolate debate in the raw food world is whether cacao is truly a nutritious raw food. The arguments range from one end of the spectrum to the other. Dr. David Klein claims in his *Living Nutrition* magazine that cacao is neither a true raw food nor is it healthy. David Wolfe, however, claims it is a raw superfood. Wolfe experimented with raw cacao extensively, and he even wrote a book dedicated to his cacao-related activities, called *Naked Chocolate*. In this chapter, we discuss the chocolate controversy, some specific perspectives on chocolate, and raw chocolate alternatives. We will also explore the cacao bean, the source of all chocolate—raw or not.

The Cacao Tree and the Cacao Bean

The cacao bean, grown on cacao trees, is the source of all chocolate. The bitterness of the food kept it from popularity in the United States until people like Milton S. Hershey found a way to soften the taste with milk and sugar. Let's look at cacao in its pure, unadulterated form. Cacao is the seed from the fruit of an Amazonian tree that was brought to Central America sometime around 1000 BC. Cacao beans were so revered by the Mayans and Aztecs that they used them as money.[1] In 1753, Swedish scientist Carl von Linnaeus named the tree theobroma cacao, which literally means "cacao, the food of the gods."

Cacao beans contain no sugar and between 12 percent and 50 percent fat depending on variety and growth conditions.[2] Raw cacao beans appear to be an excellent source of magnesium.[3] The SAD diet is low in magnesium. Although many foods contain magnesium, it is usually removed from the foods during processing. Examples of foods that contain magnesium prior to processing but none after processing include: avocados, olives, peanuts, and molasses.[4] Magnesium is a mineral that is essential for cardiovascular, digestive, neurological, musculoskeletal, and mental health. And you can get magnesium from raw cacao. It is difficult to find *official* documentation on the complete nutritional content of the raw cacao bean. However, it is possible to find official nutritional content of unsweetened cocoa powder, made from the cacao bean.

One cup of unsweetened cocoa powder contains 165 percent of the daily recommended values of manganese, 163 percent of copper, 107 percent of magnesium, 66 percent of iron, 63 percent of phosphorus, 39 percent of zinc, 37 percent of potassium, 18 percent of selenium, 12 percent of riboflavin, 11 percent of calcium, 9 percent of niacin, and 7 percent of folate.[5] Because cocoa powder is a modified form of the raw cacao bean, chances are great that the nutritional content of the cacao bean is actually higher. This information points to cacao as being a nutritious food.

Three caveats to keep in mind: First, the official nutritional information presented for unsweetened cocoa powder is for one cup of the stuff, which unless you are a full-fledged chocoholic, you are unlikely to eat during the course of a day, or even a week. Second and third are the fat and caffeine contents of the food. Depending on the cacao bean itself, the food contains anywhere from 18 to 35 percent fat content. Further, one dry cup of unsweetened cocoa powder has about the same amount of caffeine as a cup of coffee. So, the likelihood of getting a coffeelike caffeine buzz from cocoa power would only occur if you ate about a tray and a half of brownies in one sitting. It is possible, but not likely for most of us.

In the past five years, *dark chocolate* has made the superfoods list of physicians and nutritionists like Dr. Roizen and Dr. Oz. So there appears to be a mainstream recognition of the value of chocolate. There is also plenty of research to support this recommendation. Researchers have found high levels of polyphenols, which have anticancer properties, in dark chocolate.[6] They have also found that the flavanols in cocoa can boost blood flow, and suppress inflammatory responses and tendency for blood clot formation.[7] John Robbins explains, in his foreword to *Health by Chocolate,* that polyphenols inhibit the oxidation of LDL cholesterol as well as the clumping of blood cells. As a result, chocolate may protect eaters from developing atherosclerosis, or hardening of the arteries. Dr. Carl Keen and his research team at the University of California-Davis found support for the claim that ongoing consumption of cocoa may be associated with improved cardiovascular health.[8]

According to nutritionists at the University of Arizona, chocolate may be used subconsciously by some as a form of self-medication for dietary deficiencies, like magnesium or to balance low levels of neurotransmitters involved in the regulation of mood, food intake, and compulsive behaviors (serotonins and dopamine). Chocolate cravings are often episodic and fluctuate with hormonal changes just before and during the menstrual cycle, which suggests a hormonal link and confirms the assumed gender-specific nature of chocolate cravings.[9] Gabriel Cousens and David Wolfe suggest that the best strategy for overcoming chocolate addictions may be to switch from processed chocolate to cacao. In its natural state, cacao is also less likely to be allergenic, addictive, or reactive to the body.

Is Chocolate a Drug?

In his book *From Chocolate to Morphine: Everything You Need to Know About Mind-Altering Drugs,* Dr. Andrew Weil defines a drug as any substance that in small amounts produces significant changes in the body, mind, or both.[10] Generally, food is good for us. However, we have probably all experienced times when food is *not* good for us. This includes when food is abused generally or when certain substances like caffeine are abused. Widespread food abuse is evidenced by the diet book and weight-loss program megaindustry in the United States. There are many metaphors for hunger in our society. We often eat when we really need to fulfill a hunger for something else, such as increased self-esteem, a more enjoyable vocation, or better relationships. Dr. Anita Johnston describes several metaphors in her book, *Eating in the Light of the Moon,* and states, "When we can define our hungers and develop a deeper awareness of what we are hungry for, we can begin to seek the appropriate nourishment."[11]

Dr. Weil notes that most *chocoholics*, though not all, are women. He also mentions that the craving for chocolate in most women increases just before the menstrual cycle begins. Dr. Weil theorizes that chocolate is used, in this manner, as an antidepressant. He describes the case of a patient who used Alcoholics Anonymous to kick his chocolate addiction. After many unsuccessful attempts at stopping his chocolate obsession, AA proved to be a successful solution for this individual.

Whether chocolate is a drug or not depends largely on the individual and his use of the food. Of particular note is Dr. Weil's bold statement that "Human beings appear to have an inborn need for periodic variations in consciousness and drugs can be an easy route to these experiences."[12] If you are using a food, or any substance to achieve an alteration in consciousness, you are using it as a drug. And even if you are, whether this is a problem depends upon things like frequency and overall impact on your life. Depending on your orientation toward chocolate (raw or not), this information may be helpful to you in being aware of your relationship to the food, its effect on you, and its impact on your life and functioning overall.

The Raw Chocolate Controversy

Even in the face of all of the support for a more refined version of cacao beans, there are still some raw foodists who do not believe that cacao is a nutritionally sound raw food. These individuals and groups appear to be in the minority overall. And, many who previously supported the use of carob as a chocolate-like substitute are now cacao converts. If you choose to try the raw version of chocolate (unless you really enjoy dark chocolate, you will not enjoy raw chocolate), you can see how your taste buds and your body reacts.

There are several opinions on cacao and its derivatives from raw food leaders. Some of them are official, others I have taken the liberty of crafting based upon writings of these individuals. Natalia Rose includes organic dark chocolate (in particular Dagoba and Green & Black's bars) on her list of raw foods. Although I no longer eat these bars, I must admit they saved me during the transition process until I was able to find pure raw chocolate substitutes. Ann Wigmore and Norman Walker do not mention cacao or chocolate in any of their writings. Neither does Brian Clement. Although it is *raw*, cacao is not on the menu at the Hippocrates Health Institute because it is not living. Cousens includes raw cacao beans on his Phase 1 diet—foods suitable for individuals with diabetes. Raw chocolate is not addressed in his book, but at the Tree of Life's store, Cousens sells some of the best raw chocolate cups filled with nut butter and fruit that I have ever had. In addition, raw cacao sweetened with stevia would be on the Phase 1 diet. Victoria Boutenko includes a recipe in her cookbook for the *unchocolate* cake made with carob, not cacao. And Alissa Cohen uses dates and carob in all of her raw recipes, although she recently began selling raw chocolate on her Web site.

"Chocolate should be viewed as a celebratory food."[13] Jennifer Cornbleet's cookbook *Raw Food Made Easy* makes this suggestion. For anyone with a weakness for chocolate, this may be a good way to view the cacao bean and its derivatives. If you are like me, my taste buds never get sick of raw chocolate anything. However, when I overdo it, my body reacts with skin breakouts and weight gain. To combat this, I give up all chocolate during Lent. This fasting also gives me a time to rethink my position on chocolate and develop a new appreciation for it.

Carob versus Cacao

As I stated above, many of the raw cookbooks use cacao powder or raw carob chips, and many use the carob/date combination as a substitute for chocolate. The table below compares the nutritional value of equal amounts of carob and unsweetened cocoa powder. The biggest downside to carob, based on this table, is the 51 grams of sugars compared to 2 grams of sugar in unsweetened cocoa. The glycemic load[14] of carob is 31 versus 4 for cocoa. Cacao is estimated to have a glycemic load of about 1. No one can really eat unsweetened cocoa without something to temper the bitterness of the food, so this comparison can be seen as a fallacy. However, the following facts should be considered in light of the carob versus cacao comparison. First, raw cacao nibs, which are ground up cacao beans with no additives, can be eaten plain. They are not nearly as bitter as cocoa. Second, there are sweeteners used by raw foodists, described in detail in the next chapter, that can be used to keep the glycemic load quite low. These sweeteners include stevia, agave, and yacon syrup.

Table 5.1–Nutritional Content of Carob and Cocoa[15]

Topic	Carob Flour	Unsweetened Cocoa
Calories	229	197
Fat	1g	12g
Sodium	36mg	18mg
Carbohydrates	92g	47g
Fiber	41g	29g
Sugars	51g	2g
Protein	5g	17g
Calcium	36%	11%
Iron	17%	66%
Glycemic load	31	4

The other issue to consider about carob is its taste. It has a flavor all its own, which is both sweet and distinct. And like cacao, carob is ground into a powder. Carob powder is generally a little lighter in color than cacao. And although it is brown in color like chocolate, it does not really taste much like it at all. Mixed with dates, which have their own high sugar content and glycemic load, the carob is more palatable. And some recipes use avocados to temper the taste of carob. You can use the nutritional content information and your own taste buds to make a decision about whether you want to include carob in your list of nutritional "must haves."

The concept of food combining is used religiously by many raw foodists, as described in chapter 4. The use of carob and cacao together is a common food-combining practice. The reason for this is that the bitter flavor of cacao compliments the sweet flavor of carob. You can mix the two together and try them to see what you think.

What to Look for in a Good Non-Raw Chocolate Candy Bar

If you are wondering what some good criteria are for a non-raw chocolate candy bar, one that you can buy at Whole Foods or your local health food store, consider the criteria below. If you can find a brand (like Dagoba or Green & Black's) that meets all of these criteria, it will generally be better for your body than a bar with less or none of these criteria. A good rule of thumb is to use as many of the criteria listed below as you can in looking for the optimally nutritious mainstream chocolate candy bar. The bar should contain a statement on the label that says it is:

At least 70% cacao (raw cacao, if possible)
Contains certified organic ingredients
Genetically Modified Organisms (GMOs) free
Vegan (milk-free, dairy-free, also soy-free)
Raw sweetener (agave, molasses, or raw cane sugar)

Challenges in the Raw Chocolate Industry

Demand for raw chocolate usually outstrips supply. In such a situation, the science of economics would dictate that the rational supplier would raise the price for the goods until the rational consumer demand would decrease to the point where supply and demand are in sync. In reality, supply and demand are rarely exact. And all suppliers and consumers do not act rationally.

I interviewed some raw chocolatiers for this chapter and found that many have similar concerns. The chocolatiers said they generally are unwilling to raise the price of their chocolate products just because demand is higher than supply. However, there are many, many raw foodists who would willingly pay higher prices for raw chocolate. Where does this leave us in the raw chocolate scheme? It leaves us with unmet demand. A typical consumer would get angry, but that's not the raw and living food way. So the consumer in this scenario either waits or decides to make his own raw chocolate products. The process, which consists of combining raw cacao beans and agave nectar (per the goraw.com recipe), just needs to be adjusted by the consumer until he finds the proportion for his taste buds.

Other challenges in the raw chocolate industry include obtaining adequate supply of high-quality raw cacao beans. Most beans are imported from Ecuador and until recently, there were few suppliers in the United States who imported directly. Another challenge is the shelf life of the bar. Depending on the sweetener used, the shelf life can be weeks or months. Of course the typical Hershey bar or any bar with partially hydrogenated fats can have a one-year-plus shelf life. If raw chocolatiers are trying to please Americans based on past expectations, these needs will never be met. Ask yourself: Do I love the taste of raw chocolate? Do I want to eat only high-quality, *good for you* raw chocolate? If yes, then you will either have to make your own or identify several raw chocolate suppliers who can substitute for each other when one is low on product. Not all raw chocolate is created equally.

Sources of Raw Chocolate

Love Street Living Foods (www.lovestreetlivingfoods.com)
Go Raw (www.goraw.com)
Pure Food and Wine Takeaway (www.oneluckyduck.com)
Awesome Foods (www.awesomefood.com)
Tree of Life (on-site only in Patagonia, Arizona)
Café Gratitude, San Francisco and Berkeley (www.cafegratitude.com)
Gnosis Chocolate (www.gnosischocolate.com)

CHAPTER 6

Raw Sweeteners

······································

Whether you are interested in the raw food lifestyle or not, how and if your food is sweetened should be of interest to you. Synthetically produced or chemically altered products like Sweet'N Low, Equal, Splenda, or refined sugar are not in the vocabulary or the mind-set of the raw foodist. In addition, artificial sweeteners have no nutritional value and may even cause harm. Depending on the school of raw food that you subscribe to, terms like stevia (a naturally growing leaf), agave (nectar from the cactus plant), and yacon (syrup from the yacon—a root vegetable plant) may represent some of the contents of the food on your plate.

The Sugar Blues

1n 1975, William Dufty wrote the book *Sugar Blues*. The book was inspired by Dufty's withdrawal from a significant sugar addiction in the 1960s. Interestingly, Dr. Otto Warburg's finding of a causal link between sugar and cancer in 1931 (addressed in chapter 8) is not even mentioned by Dufty.

This probably has less to do with Dufty's skills as a researcher and more to do with mainstream media's ability to bury stories that create the wrong kind of controversy.

In *Sugar Blues*, Dufty describes the history of refined sugar in our diets dating back to the 16th and 17th centuries. He provides some good evidence that the known dangers of eating refined sugar date back to 1664, when diabetes was first identified by Thomas Willis, a physician in England. Willis documented the relationship between excess sugar in the urine and the likelihood of developing diabetes. But because the king's largest business at the time was importing refined sugar, Willis did not share this information publicly. A wise move for a 17th-century physician who wanted to avoid being banished. Today, we still find many of our dietary choices being driven by government policy, some of which may be to benefit industry and not the citizen.

In the preface to *Sugar Blues*, Dufty defines *sugar,* defines *the blues,* and then he defines the term *sugar blues* as the following: "Sugar is refined sucrose, $C_{12}H_{22}O_{11}$, produced by multiple-chemical processing of the juice of the sugar cane or beet and removal of all fiber and protein, which amounts to 90 percent of the natural plant. Blues is a state of depression or melancholy overlaid with fear, physical discomfort, and anxiety. And sugar blues is defined as multiple physical and mental miseries caused by human consumption of refined sugar—commonly called sucrose."[1]

Around the time that Dufty's book was published, sugar substitutes became popular. The reason for sugar substitutes was, of course, not to improve nutritional value of the food, but rather to decrease the caloric content because sugar is about 60 calories per teaspoon and Sweet'N Low is zero calories. The raw food approach recognizes both the human desire for sweetness in our food as well as the need for sound nutrition. Believe it or not, there are several natural sources of sweetness, used by many or most raw foodists, that taste wonderful and are nutritious. The one drawback (for anyone who is worried about calories) is that there is only one raw food sweetener with zero calories. That is the naturally sweet Peruvian herb, stevia.

Glycemic Load

In any discussion about sweeteners, it is necessary to address the impact of any sweetener on blood sugar. In this chapter, and throughout the rest of the book, I use glycemic load as an indicator of the food's impact on your body's blood sugar. Glycemic load, which uses glycemic index in its calculation, gives a relative indication of how much a serving of food is likely to increase your blood sugar level. As a rule of thumb, most nutritional experts consider glycemic loads below 10 to be *low* and above 20 to be *high*.[2] In the raw and living food worlds, the criteria tends to be more strict. In the case of Dr. Gabriel Cousens' nutritional program design, the glycemic index level for all food is, in general, no more than 1. A low glycemic load is good. A high glycemic load is bad.

The Raw/Living Food Sweetener Controversy

As with most philosophies, there are disagreements within the raw and living food movement. Conservatives eat only living foods and liberals eat both living and raw foods. It is important to clarify that the issue is not whether to sweeten food, but *how* to sweeten the food. From time to time, we all need something sweet. The range of philosophies on how to sweeten your food can be divided into the following seven steps.

1. Only Fresh Fruit to Sweeten

Fresh fruit provides the only way to sweeten food using *living* ingredients. Most of the other sweeteners mentioned below are considered *raw*, but not necessarily *living*. Recall from the introduction that the primary difference between these two types of foods is that living foods still contain living enzymes and raw foods do not. Drs. David Klein and Doug Graham and their followers subscribe to this living philosophy. In fact, they advocate that for most of us, a high percentage of our diet should be fruit-based. Their methodologies for sweetening food, which you can read about in most editions of the *Living Nutrition* journal, include using the following *sweet* fruits: oranges, tangerines, bananas, apples, pears, melons, grapes, papayas, berries, peaches, mangos, and similar fruits. Fresh and living fruits are very nutritious. But sugar is still sugar, regardless of the form it takes, and the natural fructose (natural sugar in fruit) content in most fresh fruits can carry a glycemic load from 4 to about 12 per serving.

2. Fresh and Dried Fruit to Sweeten

In addition to the above list, living food advocates also support the use of dried fruit for sweetening purposes. The most common dried fruits used for sweetening are dates, figs, raisins, and goji berries. These fruits should be dried without any additives or chemicals. You can find organic fruit in health food stores, but most supermarkets sell dried fruits with preservatives and other chemicals that can wreak havoc on your digestive system. The glycemic load for dried fruit is quite high and can range anywhere from about 35 (for dates) to 44 (for figs) and 75 (for raisins). Caloric content for dried fruit is high, and it ranges from about 150 to 300 calories per serving.

3. Coconut Pulp, Zucchini, and Stevia

Dr. Gabriel Cousens has designed an intricate system for determining foods that are processed most efficiently by the body, whether you are disease-free or have a chronic disorder like diabetes. His Phase 1 diet is the food plan he considers to be the ideal. It includes only coconut pulp, zucchini, and stevia as sweeteners. The glycemic load of these sweeteners is 1 or less. You can find out more about Dr. Cousens' dietary recommendations in his book *Rainbow Green-Live Food Cuisine*. Although you may not have thought of zucchini as a sweetener, when it is shredded and added to recipes, zucchini adds a good, natural, light sweetness.

4. Coconut Water, Mesquite, and Maca

These sweeteners reflect the next (or more liberal) level of Dr. Cousens' diet, which he calls the Phase 1.5 diet, or not quite the ideal. In addition to coconut pulp, stevia, and zucchini, Dr. Cousens' Phase 1.5 diet includes coconut water, mesquite, and maca.

5. Sweet Vegetables (Yams, Pumpkin, Beets, Carrots), Carob, and Bee Pollen

These sweeteners are included in the next level of Dr. Cousens' Live Food Cuisine diet. This is what he calls the Phase 2 diet, and it is meant to be a transitional phase on the way to the ideal 1 or 1.5 dietary level. Individuals may spend time in each of the three levels, depending on their specific needs at the time. The only exceptions to this would be for individuals with diabetes. For those individuals, Dr. Cousens has written an entire book entitled *There is a Cure for Diabetes*, which addresses most nutritional questions anyone with diabetes may have.

6. Raw Agave Nectar and Yacon Syrup

Raw food converts from the mainstream tend to get stuck in the *agave nectar area* when it comes to sweeteners. Even though the nectar is about 80 percent fructose, the glycemic load is very low at 1. Some raw foodists do not believe that it is possible to produce a pure raw version of agave because it must be heated to extract it from the plant. However, there are manufacturers who claim that the agave is raw, and that the syrup was not heated above 115° F. Yacon is a relative newcomer to the raw food world, but it has been recommended for use by those with diabetes. Agave in particular is probably the most palatable raw sweetener, although at 60 calories per tablespoon, it is fairly high in calories.

7. Raw Honey

There are raw food cookbooks and raw food restaurants that use honey in their recipes, probably because they believe the nutritional benefits outweigh philosophical issues. Although honey is not vegan, there are many raw food stores that sell raw honey. Raw honey is full of nutrients, although the glycemic load and calorie content are both fairly high at 10 and 65 calories per tablespoon, respectively. The lingering problem is that honey is a product of the bee, a living thing. Therefore, it violates the vegan principles.

Raw Sweeteners in Alphabetical Order

The definitions provided below contain information about the origin of the sweetener, a description of it, basic nutritional information, its use in food preparation, and where it can be purchased.

Agave

Agave syrup, sometimes called agave nectar, is a sweetener from Mexico and the southwestern United States. The substance can be obtained from several species of the agave plant, including *Agave tequilana* (also called Blue Agave or Tequila Agave). Based on the plant of origin, agave is either amber (dark) or light in color. Although official nutritional data are not available, instructors at the Tree of Life and others believe that the darker agave has a lower glycemic load than the light agave. Since they taste pretty much the same, it might be a good idea to choose the amber agave if you are going to use it.

Agave syrup is sweeter than honey, though less gummy. Uncooked agave syrup is about 80 percent fructose with a glycemic load of 1 that triples to 3 (still relatively low) if cooked. A tablespoon of agave is about 60 calories. The substance contains some calcium, magnesium, and iron.

Agave syrup can be substituted for sugar in recipes. Generally, use one-third cup of agave syrup for every one cup of sugar in the original recipe. The quantity of liquids in the original recipe should also be reduced due to the moisture included in the syrup. If you are using the agave in a cooked recipe, reduce the oven temperature by 25° F. Agave can be purchased in Whole Foods stores, most health food stores, and even online through Amazon.com. Purchasing a brand like Madhava, Wholesome Sweeteners, or Natural Zing will ensure that the agave is *raw*. Unless the word *raw* is stated on the label, you should assume the agave is cooked, which at a minimum, will increase the glycemic load from 1 to 3 and likely decrease nutrient value. The average price for a 16-ounce bottle of agave is about 10 dollars.

Bee Pollen

Bee pollen is brought to us by the female worker bee. The pollen from flowers sticks to her legs and then as she returns to the hive, she passes through a man-made screen, where the pollen rubs off and is collected. The high life force in pollen comes from the living plant energy contained within it as a result of the female bee's collection process.[3] Bee pollen is another religious food mentioned 68 times in the Bible and is also noted in the Talmud and the Koran. Bee pollen contains most naturally occurring vitamins, minerals, carotenoids, bioflavonoids, amino acids (for protein building), fatty acids, and enzymes. Caloric content is about 20 calories per tablespoon and glycemic index is zero. You must be very careful purchasing and handling bee pollen as it can become rancid if not stored properly. Purchase bee pollen from a reputable local beekeeper or from a natural-food source who receives the pollen fresh and immediately freezes it. You should keep the pollen frozen as well.

Bee pollen tastes vaguely like honey with a slightly medicinal taste. When used in certain recipes, this can be masked using other foods. More importantly, bee pollen is seen as a raw and living superfood and is supported by most nutritionists and physicians who are interested in nutrition. For example, Dr. Cousens says that bee pollen is "the proactive life force of the plant world . . . the finest food and best medicine ever discovered. Pollen contains the richest source yet of vitamins, minerals, proteins, amino acids, hormones, enzymes, and fats."[4] As a supplement, you can take about a tablespoon of bee pollen a day for maximum benefit. A negative side effect of bee pollen is for those who are allergic to bees. If you are allergic, it is best to avoid bee pollen. While local, fresh purchase of bee pollen is probably your best bet, you can purchase raw bee pollen online at places like Natural Zing (www.naturalzing.com) for about 10 dollars per pound.

Carob

Carob is a member of the legume family, and as such, its roots host bacteria that convert atmospheric nitrogen into nitrates, which can be used by plants to make proteins. Carob beans are grown in pods and then crushed and used as a food substance. It is another religious food, traditionally eaten on the Jewish holiday of Tu Bishvat. Carob juice drinks are traditionally drunk on the Islamic holiday of Ramadan. Carob pods were the most important source of sugar before sugarcane and sugar beets became widely available. Carob powder and carob chips are sometimes used as an ingredient in cakes and cookies instead of chocolate.

A cup of carob flour (the only version of carob officially titrated in nutritional data) is 229 calories with 51 grams of sugar. It has only one gram of fat, which is why it is preferred by some to chocolate. And before raw, vegan chocolate was available, carob could be used as a vegan chocolate substitute. It does contain a fair number of minerals including calcium, magnesium, potassium, and manganese.

Raw foodists eat pure carob powder, not carob flour, which has other additives. One of the challenges with carob is finding a truly raw version. Most carob powder is at least slightly roasted. Natural Zing does carry a truly raw version, imported from Spain, for about seven dollars a pound.

Coconut (Pulp and Water)

The type of coconut we are talking about here is the young Thai coconut, not the round, dark brown, hairy versions. Young Thai coconuts are white on the outside and pointed at the top. There are several caveats to keep in mind if you want to obtain the maximum benefit from the coconut meat and water. The coconuts should be organic, never soaked in formaldehyde (as they sometimes are to preserve them during travel), and as young as possible. If you are not immediately familiar with the young Thai coconut, you have probably seen them in Whole Foods or other markets. There are a few reliable Internet resources for the coconuts as well, like Genefit Nutrition (www.genefitnutrition.com). But because availability is tenuous, you should always verify with each order that the coconuts are organic and formaldehyde-free. Let's first look at coconut water and then coconut pulp.

Coconut water has a low glycemic index, but it is still higher than the glycemic index of coconut pulp. The coconut acts as a natural water filter that takes almost nine months to filter each liter of water. The water travels through many fibers, being purified, where it is stored away sterile in the nut itself. This coconut water is pure and clear and is one of the highest sources of electrolytes known to man. *Coconut water is identical to human blood plasma, which makes it the universal donor.* Plasma makes up 55 percent of human blood. Coconut water was used to transfuse wounded soldiers during World War II.

You can obtain coconut water by using a cleaver to loosen up the coconut's casing. Once you have done that, you can pour the water from the coconut. It is best to drink the water immediately, but you can keep it refrigerated for a few days and it will retain most of its valuable properties. In addition, you can use coconut water in recipes that call for fluids and sweeteners. You will have to experiment with specific proportions during the food preparation process to get the flavor you want.

Coconut pulp is on Gabriel Cousens' Phase 1, or optimal raw diet. It has a glycemic index of 1 or less. Coconut pulp, which lines the inside of the coconut casing, can be scooped out with a spoon. This pulp is white, sweet, and has the natural consistency of a thick gelatin. It tastes amazing on its own, and it is used by some of the most renowned raw food leaders, like Sarma Melngailis, to create raw frozen ice cream and other tasty nutritional delights. When my husband, Joe, first stopped in Pure Food and Wine on his own for dinner, he ordered their classic ice cream sundae for dessert. Joe was so pleased with the dish that he could not believe it was actually *good for him.* He called me several times while he was eating the sundae to confirm and reconfirm that what he was eating was, in fact, good for him. Joe still returns whenever he can for dinner and a classic sundae.

Goji Berries

Goji berries are native to Tibet and the Himalayan Mountains. They are a sweet, reddish, dried fruit slightly under the size of a raisin. The goji berry contains 18 amino acids, more beta carotene than carrots, vitamins B_1, B_2, and B_6, and E, and 500 times the amount of vitamin C that oranges have. According to Gabriel Cousens, goji berries have been found to be effective in increasing white blood cells, lowering cholesterol, and burning body fat.[5]

One-quarter cup of goji berries contains 90 calories with 12 grams of sugar, 4 grams of protein, and 24 grams of carbohydrates.

Goji berries also come in juice form, but the juices are usually high in added sugars as they are created from a concentrate. It is best to eat goji berries dried. And if they are too chewy or dry, soak them in water for 30 minutes before eating them or using them in a recipe. You can buy goji berries at Whole Foods, health food stores, and many online stores for an average price of about 18 to 20 dollars per pound.

Honey (Raw)

The term *vegan* was coined by Donald Watson in 1944 as "a way of living that excludes all forms of exploitation of and cruelty to the animal kingdom and includes a reverence for life." It applies to the practice of living on the products of the plant kingdom to the exclusion of flesh, fish, fowl, eggs, honey, animal milk and its derivatives, and encourages the use of alternatives for all commodities derived wholly or in part from animals.[6] Honey is made when flower nectar is mixed with bees' saliva, hence the vegan issue. Because it is sold and used by some raw foodists, I address it in this chapter. Again, based on your own assessment, your beliefs, and the needs of your body, you will need to determine what path is the right one for you.

Because the processing of raw honey removes valuable nutrients from the substance, we will only address honey in its raw form here. Raw honey is probably not the honey you are used to. In its raw state, honey takes on a harder texture and is whitish in color. The pollen that collects on the bees' legs as they move from plant to plant is only as healthful as those plants. Some phytonutrients found in honey have been shown to possess cancer-preventing and antitumor properties. Researchers have discovered that these substances prevent colon cancer in animals by shutting down activity of two enzymes.[7]

Honey increases calcium absorption, can increase hemoglobin count and treat or prevent anemia caused by nutritional factors, fights colds and respiratory infections, and provides an array of vitamins and minerals. Raw honey is exceptionally effective internally against bacteria and parasites. Plus, raw honey contains natural antibiotics, which help kill microbes directly.

Raw honey, when applied topically, speeds the healing of tissues damaged by infection and/or trauma. It contains vitamins, minerals, and enzymes, as well as sugars, all of which aid in the healing of wounds.[8]

Raw honey is about 80 calories per ounce and has a glycemic load of 14. It has the same sweet flavor as its cooked counterpart. Unless you plan to dehydrate it or use it in a blender or food processor, you may need to place it in a warm location to soften it before use. You can buy raw honey in health food stores and on the Web for about 12 dollars a pound.

Maca

Maca is an off-white powder that has a vaguely sweet taste to it. Maca root grows in the mountains of Peru. It is a radishlike root vegetable that is related to the potato family. Maca contains significant amounts of amino acids, carbohydrates, minerals, including calcium, phosphorus, zinc, magnesium, and iron, and vitamins B_1, B_2, B_{12}, C, and E. Peruvian maca also includes a number of glycosides (sugar molecules that help remove toxic waste from the body). Official nutritional data for calories and glycemic load for maca is unavailable at this time. Maca has been claimed to have significant immunoprotective properties and to aid in hormone function, among other benefits. It is commonly added to smoothies and other drinks as a sweetener, but also for its nutritional benefits. Maca has a very distinctive flavor, so the best strategy is to taste test the powder both plain and mixed into a small amount of smoothie or other drink to determine whether it is to your liking. You will not need much powder to flavor your drink. Maca can be found in most health food stores and online for about 18 dollars for a seven-ounce jar.

Mesquite

Raw mesquite is a tan-orange, sweet, and pungent-smelling powder. Mesquite bushes grow pods. Mesquite powder is then ground from the mesquite pods. Mesquite powder is high in protein, calcium, magnesium, potassium, iron, and zinc. It has a sweet taste, similar to molasses. Official nutritional data and exact glycemic load for mesquite is unavailable at this time. Because its texture is similar to maca, you can use it in the same way, in drinks and smoothies. However, I have added mesquite to breads, crackers, and pies. You will find mesquite included in some raw food recipes as well. It has a distinct taste and like maca, you should test it before you use it in a recipe or a drink. You can buy mesquite powder on the Web for about 10 dollars a pound.

Stevia

The herb stevia, has been used both nutritionally and medicinally for centuries by the Guarani Indians of Paraguay. By the 1970s, Japan started using stevia commercially and today, they are the biggest users of the extract, which has captured 50 percent of Japan's sweetener industry. Today stevia is grown around the world from China, Japan, and other Asian countries to South America, Europe, India, the Ukraine, and even North America.

Stevia's most obvious and notable characteristic is its sweet taste. It is a great sweetener alternative because it does not raise blood sugar levels. In fact, research has shown that concentrate of the leaf has a regulating effect on the pancreas and helps stabilize blood sugar levels. Stevia is useful to people with diabetes, hypoglycemia, and Candida infection. However, it does have an aftertaste similar to artificial sweeteners, so you will need to do a taste test before assuming this is the solution for you.

Stevia can be found in four forms. The fresh leaves are the most pure form, but they are difficult to find. Dried leaves are used in brewing herbal teas and for making liquid extracts. The powder form is dried leaves ground into a fine powder. It is used in teas and cooking but does not dissolve well. The powder form is usually about 10 to 15 times sweeter than sugar.

Stevia can also be found in liquid extract form. The extract is a concentrated syrup derived from the dried leaves. Using the version made with water and not alcohol will ensure you are getting the purest form of the plant. Liquid stevia made without alcohol is dark green in color. You can buy stevia in its dry or liquid form in most health food stores and on the Internet. Because stevia is not mass-produced like the common sugar substitutes on the market, you will likely pay more for stevia than Equal or Splenda. For example, you can purchase Splenda or Equal on Amazon.com for about 2.5 to 3 cents per packet, whereas the same amount of stevia is priced at an average of 7 to 8 cents per packet.

Yacon Syrup

Yacon, which is grown in the Peruvian Andes, is a sweet tuberous root vegetable that produces a rich, sweet, brown syrup. Because it has a negligible glycemic load, it has been used successfully in the diets of individuals with diabetes. Yacon is naturally low in calories, and it contains about 30 calories per tablespoon. Yacon is also believed to promote beneficial bacteria in the colon. Research has proven that yacon is beneficial for those with hypertension. By thinning the blood, yacon can lower blood pressure by 5 to 10 percent. It can also lower cholesterol and discourage clot formation.[9]

Yacon is high in oligofructose, a naturally occurring sugar, which the human body does not metabolize, hence its potential use for diabetics and in body-weight control. Yacon resembles maple syrup in both look and taste. The cost is about 11 dollars for a 13-ounce bottle. You will probably use less yacon than agave to obtain the same level of sweetness in your food.

CHAPTER 7

Water

· ·

Is your drinking water dead or alive? There has been so much research performed on drinking water in the past two decades that it is worth reading about the distinction between dead and living water. One way to ensure you get sufficient living water is to eat or drink many fresh, uncooked, organic fruits and vegetables every day. Fruits and vegetables provide the purest water because the plants, by using their own distillation process, have done the filtering of the water for you. Scientists and researchers have spent considerable time in the past decade defining living water, determining how we can get it, and linking it all up to quantum theory.

Two things define water: structure and content. There are two types of structure. First is the molecular structure of water which we know to be H_2O. The second is its physical or crystal structure. When it comes to water's impact on your body, its crystal structure may be just as important as its molecular structure. Author and scientist Dr. Masuru Emoto began studying the crystal formation of water by photographing the crystals during the freezing and thawing process. Similar to patterns found in snowflakes, Dr. Emoto found that the crystals were never exactly the same. Sometimes the water crystals were almost perfect, while others were deformed, and sometimes, no crystals formed at all. A pattern started to arise in his research, and Dr. Emoto found that "different water formed different crystals."[1] He then found that certain things, like adding chlorine to tap water, would destroy the natural crystal structure of the water. Dr. Emoto's book, *The Hidden Messages in Water,* which includes color photographs of water crystals taken under certain conditions and in certain places, is worth a look! You may find it a challenge to believe some of Emoto's claims. For example, he documents that water crystals remained perfectly intact when exposed to kind words like *love* and *gratitude* and became deformed when exposed to phrases like *you fool* and *you make me sick.* However, much of Emoto's research finds support in quantum theory.

What is in your drinking water? Research has been performed on drinking water that contains additives such as minerals, which may not always be beneficial. Sometimes additives destroy the natural pH of the water. More importantly, additives like fluoride, pollutants, or even distilled water corrupt the natural crystal structure of water. The body is unable to process unstructured water as efficiently as it can process structured water. Scientists like Emoto, Batmanghelidj, Hendel, and Ferreira concur that when you change the structure of water, it brings little benefit to the body. Gabriel Cousens and other raw food leaders also subscribe to this belief. The belief is analogous to the principle of raw food theory that says you must eat food in its raw, uncooked state. Cousens corroborates this by stating that the more structured your drinking water, the easier it is for enzymes to digest and for vitamins and minerals to be assimilated into your cells.[2] Additional studies found the intracellular water of cancer cells to be much less structured than normal healthy cells.[3] The more natural and untouched the substances are that we put into our body, the more benefits we receive. The reverse can also be hypothesized: the more unnatural and refined substances we put into our body, the more harm that will come to it.

Preserving the crystal structure of water raises two essential questions. The first is how do we, or can we, clearly identify what types of actions destroy the crystal structure of water? The second is, given the damage to our earth over the past few centuries, most of it man-made, how do we ever get back to a purely natural water source? As it turns out, the answer to both of these questions is not easy or simple. And, when it comes to the solution, everyone seems to have a version of what we should do. Let's analyze the information and you can determine the best workable solution for you.

The information available about drinking water is abundant but often misleading. We need to be cognizant of marketing ploys and conflicts of interest that can influence writers and researchers. I tried in my research to identify conflicts of interest in information about water that seemed otherwise sound. If I did identify a possible conflict, I included it in the analysis if I was able to corroborate the recommendation with another source. In addition to content and structure, there are two additional properties of water that were mentioned by most researchers: source and pH level. Pure water (or natural water without additives) has a pH value of 7.0, making it neutral—meaning that it does not produce acid or alkaline effects in the body when consumed.[4] While the preferred source and pH level were not consistent among all writers, these categories were always included in analyses.

I have provided below a description of the three primary sources of water, natural, tap, and bottled, as well as the recommendations from several experts regarding what, if anything, you should do before drinking water from each of these sources. There are varying viewpoints, but some arguments are stronger than others, as you will see. I have broken down the details within the different philosophies into information that will be easy for you to apply. I have provided references for each section so if you are interested in investigating any or all of the information further, you can do so.

Carbonated and Distilled Water

Neither carbonated water nor distilled water is addressed below. Carbonated water (water with bubbles) is excluded because every one of the physicians and scientists who addressed it agreed that carbon dioxide in drinking water makes carbonic acid, and this makes our bodies more acidic. Distilled water has virtually all of its impurities removed by boiling the water and then condensing the steam into a clean container, leaving solid contaminants behind. Distilled water is excluded because the experts agree that distilled water is dead, unstructured water. It is so foreign to the body that you actually get a temporary high white blood cell count in response to drinking it.[5]

Tap Water

Tap water in many parts of the world is unhealthy and in some places, undrinkable. There are a few exceptions. For example, in Emoto's book, *The Hidden Messages in Water,* he studied the crystal formation in the tap water of cities throughout the world in cities like Tokyo, Paris, London, Rome, Sydney, Bangkok, and Hong Kong. He found that the water did not form crystals at all, meaning there was no structure to the water. However in Washington DC and New York City, Emoto found the water formed perfect crystals. He hypothesized that this might be the result of extra efforts taken in certain cities to protect the public water supply. For example, in New York City, special cedar tanks are used to store water.

According to Hendel and Ferreira, there are several reasons that American tap water, in general, is unhealthy. These include the fact that up to 300 pesticides and fungicides pollute the water. Also, the pressure in water pipes destroys the structure of the water.

The processing of choice for water, according to Hendel, Ferreira, Cousens, and Emoto is a three-step process that includes reverse osmosis (ROI), a carbon filter, and revitalization by soaking crystals in the water for eight hours. The ROI unit removes bacteria, viruses, nitrates, fluorides, sodium, chlorine, particulate matter, heavy metals, asbestos, organic chemicals, and dissolved minerals. The carbon filter removes organic and inorganic chemicals that the ROI unit cannot. The hexagonal structure of the quartz crystals will rearrange the water to its original structure. Crystals have their own particular vibration of a precise and measurable intensity. This vibration attunes itself to human vibration better than any other gem or mineral. It can also attune itself to the vibration of substances, like water. Quartz crystals are used to amplify, clarify, and store energy. Quartz has long been recognized for its ability to produce electrical impulses. Pressure on quartz crystal generates a minute electrical charge called piezoelectricity.[6] These characteristics are all involved in the process of reorganizing water into its own crystal-like structure.

Batmanghelidj, like Cousens and Hendel, is a medical doctor, but he has a different approach to tap water. He believes that the most important issue is drinking water regardless of its makeup. In his book, *You're Not Sick, You're Thirsty,* Batmanghelidj presents some good evidence that most Americans suffer from dehydration, and we could significantly decrease healthcare costs if we all just drank more water, regardless of where it comes from. Batmanghelidj does suggest, if you have a concern about toxins, to either allow your tap water to sit out for an hour (to allow chemicals like chlorine to evaporate) or use a carbon filter on the tap.

Depending on where you live, your choices for action regarding your tap water are varied. In New York City, you can choose to just drink the water directly. Or you can take the laid-back Batmanghelidj approach and let your water sit out for an hour before you drink it. On the other end of the spectrum, you can proactively purchase a reverse osmosis unit and a carbon filter and use crystals to revitalize the structure of the water that was taken apart, at least partially, by the ROI and the filter.

Natural Water

The two primary sources of natural drinking water are spring water and artesian well water. Spring water comes from a natural spring. It bubbles up to the surface and along its path, it can attract impurities from the ground and the air. Once it reaches the surface, there is nothing to protect it from being exposed to bacterial growth, ground pollutants, or other contamination. In order to obtain the maximum benefit from spring water you need to do two things: (1) allow it to mature underground completely so it achieves its full structure and (2) bottle it at the source so it is pure and not touched by surface pollutants or bacteria. This process has implications for everyday life. First, most of the spring water found in supermarkets is immature because of man-made intervention to bottle the water. Second, many of these waters are ozonated (water that has had unstable oxygen molecules added to it), which destroys the frequency pattern, and can be considered dead because they have lost most of their structure. [7] Third, in order to obtain full benefit from spring water, it really needs to be available in your local area in a place where you can go directly to fill up your bottle. Most pure sources of water like that described by the water experts are likely bottled, mineralized, and distributed through supermarkets or convenience stores.

Artesian well water may be the best water for us because it is fully developed and surfaces on its own without any man-made intervention. Artesian water comes from a source deep within the earth that is protected by a solid confining layer made of stone and/or clay. In the case of the artesian water from islands like Fiji, natural artesian pressure forces the water through a completely sealed delivery system free of human contact. It is never exposed to the environment. According to the Environmental Protection Agency (EPA), water from artesian aquifers often is more pure because the confining layers of rock and clay help to provide a protective shield from potential contamination. People who have the good fortune to live near artesian wells may have the best option in water. But what about the rest of us?

Bottled Water

Bottled water is probably the most common way that most of us obtain our water today. The issues around bottled water are threefold. First, you need to consider the source. Second, you need to consider any added content. Lastly, but perhaps most importantly, you need to consider the impact of plastic bottles on your body and the environment. Let's look at each of these issues individually.

Consider the source. Much bottled water today is simply *purified* tap water. It is likely not purified using reverse osmosis, carbon filtering, and crystal revitalization. The least expensive form of filtering is probably most common. And the end result may be nothing very different from your own tap water. In some cases, the bottled water may actually be less healthy than your own tap water. Read the label to determine the source. If the label does not say *source* or *spring,* it is likely filtered tap water. Often the label just says *drinking water.* Some minerals may be added during the process, but often our bodies are unable to process and use these minerals because of size and amounts used.

Added content. If your bottled water is spring water and it is bottled at the source there is still a chance that it is immature and the structure is not complete, but the quality of the water will be superior to bottled tap water. All of the water experts, Cousens, Hendel, Ferreira, and Emoto, agree that bottled water from a mature spring or artesian well is the best bet for both healthy structure and content. One of the brands mentioned by Hendel and Ferreira is Fiji water. With the exception of the plastic bottle (addressed below), because it comes from a natural artesian well on the nonindustrialized island of Fiji, Fiji water appears to be the superior bottled drinking water.

Impact of plastic on the environment. The last topic to be addressed regarding bottled water is the bottle in which the water is contained. Plastics can break down. The molecules from the plastic seep into the water and can impact the purity and healthfulness of the water. While claims have been made about this and it seems to make perfect sense, no hard evidence has ever been presented. The reasons for this are pretty obvious. The solution would be to purchase only water that is bottled in glass, which can be cost prohibitive. The second issue regarding the bottle is the impact on the environment. Ideally, we do not want to be buying and throwing away a plastic bottle every time we need a drink of water. Even if the bottles are recyclable, the amount of energy it takes to recycle and the resulting landfill overflow is astounding.

The consensus regarding bottled water seems to be: (1) Don't drink it unless you have to. The problem is, unless we can obtain suitable tap water, many of us may feel we have no other choice. (2) If you have to drink bottled water, buy it in glass bottles, if possible. (3) Drink water that contains the term *artesian* or *spring, bottled at the source* on it. Finally, as noted above, only drink flat water, not carbonated.

"We should look at water as our primary most important food source that replenishes us and nourishes us with its necessary living energy and information."[8]

pH Level

· ·

Did you know that in 1931 Dr. Otto Warburg received the Nobel Prize for identifying a causal link between sugar and cancer of all types? While we can only speculate why his work, which continued until the 1980s, has not been splashed across the headlines of every national and local newspaper, we can still benefit from his findings by applying the principles in our own lives. Because sugar is a highly acidic food, Dr. Warburg's findings, described in detail later in the chapter, have everything to do with the acid-base (alkaline) balance of the body, also known as our pH level.

The term pH actually means *potential of hydrogen* or, the number of hydrogen ions in a substance. The higher the number, the more acidic the substance. The lower the number, the more alkaline the substance. A pH of 7 is neutral, neither acidic nor alkaline. In order to remain healthy, the body must maintain an overall alkaline pH in the range of 7.36 to 7.45. A balanced diet is generally composed of about 75 to 80 percent alkaline foods and 20 to

25 percent acidic foods. Alkaline foods include fruits, vegetables, greens, and sea vegetables. Acidic foods include all sweets, grains, meats, fish, eggs, milk products, animal fats, coffee, tea, unripe fruits, and alcohol.[1] Acidic foods, like sugar, contain little to no oxygen.

The food you eat determines your pH level. But the food your body needs and how your body processes the food is determined by your uniqueness, or your biochemical individuality (discussed in chapter 4). In his book *Conscious Eating,* Gabriel Cousens describes different constitutional types. He describes how, depending on your dominant constitutional (or body) type, your body may process acid and alkaline foods differently. There are a series of questions that he has designed around the constitutional types. Based on your responses, you can determine your dominant type and this should give you additional insight into how your body is likely to process the foods you eat. Similar to the concepts of biochemical individuality, what Cousens stresses is important: we are all unique, and we must first understand this and then observe and apply what we learn about ourselves and our nutritional habits.[2]

What we eat plays a critical role in the pH balance of our bodies. Renowned nutritionists like Paavo Airola believe that excess acidity in our bodies is a basic cause of all disease. In addition to inflammatory diseases and infections, a person with an acidic pH also will experience dulled mentality, slower thinking, headaches, and depression.[3] A more acidic pH also causes irritability because of the loss of calcium, magnesium, and potassium from the cells. Dr. Susan E. Brown does an excellent job of analogizing our treatment of the environment to our treatment of our bodies. She says, "Just as our planet is undergoing the process of demineralization of its soils and waters due to the acid products of fossil fuel consumption, we are suffering the demineralization of our bodies due to our acidic dietary practices."[4]

If you are interested in self-testing your urine for pH levels, please see either Gabriel Cousens' book, *Conscious Eating* or Susan Brown's book, *The Acid Alkaline Food Guide.* Both describe the process for testing your urine in a reliable way. Your other option is to work directly with your primary care physician, who can add value in the interpretation of the results and can conduct tests that are more detailed and more reliable.

pH and the Raw Food Lifestyle

While these pH facts are interesting, you may be wondering what all of this has to do with the raw or living food lifestyle. For one thing, the more acidic we get, the less life force energy we have in our bodies, because we are taking in less oxygen. In addition to foods, some studies have shown that fluorescent lighting, radiation from computer screens, and wireless technology also compromise our ideal pH balance. In her book, *Raw Food Life Force Energy*, Natalia Rose makes several suggestions regarding how we can combat activities that imbalance our energy. She recommends that you (1) spend as much time outdoors in sunshine as possible, (2) breathe deeply to oxygenate your energy levels, (3) practice yoga, dance, or get a massage, and (4) keep negative emotions in check.[5] On Rose's list of highly alkaline (good for you) intake is: sunlight, fresh mountain air, fresh green vegetable juice, raw vegetables, and sprouts. She also recommends the use of colonics (addressed in the chapter on healthy conflict), natural bristle body brushes on dry skin, infrared saunas, and Chi machines.

Chi is the Chinese word that refers to *life force* or *life energy*. The Chi machine is a small machine that you place under your ankles while lying down. The machine vibrates and moves to the left and right for 15-minute intervals. The machines are supposed to oxygenate, strengthen the body and the mind, increase positive energy, stimulate the lymphatic system, and balance chakras. A chakra is a spinning sphere of bio-energetic activity from the head to the base of the spine.

Raw foods have an ability to restore and maintain the pH balance of our cells. Some of the reason for this is simply that food, in its uncooked state, is less acidic. Some of the most famous raw and living food experts advocate highly alkaline diets, among them Norman Walker and Ann Wigmore. Dr. Walker's daily diet consisted primarily of vegetable juices, most of which were green. Dr. Wigmore's primary source of nutrition was wheatgrass juice, a highly alkaline substance. In addition to fresh vegetables, she ate sprouts daily, which are also a high source of alkalinity. Dr. Wigmore's health institutes continue to carry out her work today bringing health through these same alkaline focused nutritional practices. In Victoria Boutenko's book, *Green for Life,* she makes an excellent case for the value of the *green smoothie.* Drinking greens, even if they are mixed with bananas or other fruits, is another example of raw food supporting a more alkaline balance. Brian Clement's menu at the Hippocrates Health Institute in West Palm Beach, Florida is yet another example of the raw food diet, a highly alkaline diet. Using fresh vegetable juices and primarily fresh vegetables, the institute has brought many cancer victims from a state of disease to one of healthfulness.

Cancer and Acidity

In 1931, Otto Warburg won the Nobel Prize when he demonstrated that cancer cells thrive in conditions that are without oxygen. Acidic foods contain little to no oxygen. Warburg hypothesized that cancer was a defect in how a body breaks down glucose. In a 1966 speech to the Nobel Laureates, he said that, "The prime cause of cancer is the replacement of the respiration of oxygen (oxidation of sugar) in normal body cells by fermentation of sugar."[6] You can read Warburg's Nobel Lecture, delivered on December 10, 1931, entitled "The Oxygen-Transferring Ferment of Respiration" on the Nobel Prize Web site at www.nobelprize.org. A very simplified application of Dr. Warburg's principles for avoiding cancer would be the following: (1) avoid acidic (low oxygen) foods—or, in the alternative, eat more alkaline foods, (2) take in plenty of oxygen and fresh, clean air, and (3) do not eat sugar or anything containing sugar. Rose's rules for keeping in balance are actually an updated version of Warburg's. So, almost 80 years after his Nobel Prize, Warburg's suggestions have stood the test of time.

According to Warburg, the most important of these is not eating sugar. If, minimally, you can stop eating refined sugar or anything containing refined sugar, it would be a step in the right direction. Dr. Warburg did not experiment on naturally occurring sweetening substances like agave and yacon, so it is unclear whether these foods would have the same impact as sugar. Although we all need some sweetness in our diets, it is probably best to keep the sweeteners with a glucose base (like agave and yacon) to a minimum. On the other hand, stevia has a neutral pH of 7, so it may be a safer long-term strategy—if you like the taste. Even if you are not on a living and raw food path, all three of these practices—low acid foods, a lot of fresh air, and no sugar—should become part of your daily habits.

Cooking and Acidity

Cooking food makes it more acidic. When you eat a diet that is made up of raw fruits, vegetables, sprouted nuts, seeds, and grains, you will begin to shift your balance to a more alkaline state.[7] In the beginning of the chapter, we saw that the ideal diet is made of 75 to 80 percent alkaline foods and 20 to 25 percent acidic foods. One way to successfully translate this into a healthy raw food diet is to intake 75 to 80 percent fresh vegetables and fruits (with a good amount of leafy greens, fresh or juiced) and 20 to 25 percent from nuts, seeds, and grains. Sprouting your nuts will make them less acidic as well.

Common Highly Alkaline and Highly Acidic Foods

The following is a list of foods that most nutritional experts, whether they are raw foodists or not, agree are highly alkaline producing. There are many other foods that are alkaline producing overall. The ones on the list are only those felt to be most alkaline producing, so they produce the greatest amount of value for you nutritionally. You can find a more comprehensive list of acid to alkaline foods in *The Acid Alkaline Food Guide*. While you can certainly (and should) eat foods other than these, any food on this list will help in getting you back into balance if you have been eating too many acidic foods. In general, for the foods on the list below, higher alkalinity will result if the foods are not cooked, though even cooked, all are considered to be excellent sources of alkalinity.

Asparagus (may be the best according to most experts!)

Celery

Chestnuts

Collard greens

Dulse

Endive

Figs

Ginger root

Kale

Kelp

Lemon (produces alkalinity even though it tastes acidic)

Lime (same as lemons)

Onions

Pumpkin seeds

Seaweed

Sweet potatoes

Wheatgrass juice

Yams

The following is a list of foods considered to be highly acidic. In general, these should be eaten in very small quantities, and if possible, avoided altogether. *Note that none of these foods is included on a raw or living food diet.*

Bacon

Beef

Cheeses, especially aged

Chocolate milk

Cocoa powder (this is not the raw cacao bean discussed in chapter 5)

Egg noodles

Expresso

Fried foods

Jams and jellies

Ice cream

Iodized table salt

Lobster

Macaroni and cheese

Milk chocolate

Pizza

Pasta with white flour

Red wine vinegar

Refined sugar

Refined white flour products of all types (bagels, biscuits, bread,
	donuts, pancakes)

Salami

Shellfish

Shrimp

Soda

Soy beans

What's Your Best Strategy?

Whether you are reading this book to become a raw foodist or just to learn more about the living food lifestyle, the information in this chapter may persuade you that eating fresh, raw, organic food may just be a good idea after all. Part of your strategy could include thinking about what you will do with all of the extra energy you have once you stop eating sugar and add greens to your daily intake. And what about the extra time you will have? Because after all, your meals will be quite simple, and you will only need to shop in the produce aisle of the grocery store.

CHAPTER 9

Healthy Conflict

· ·

Differences of opinion that take the form of expressed, healthy conflict are good. When people who subscribe to the same overall philosophy disagree, the conflict will generally lead to a revalidation of the original path or a change in thinking at some point. The necessary elements for a positive outcome of healthy conflict in any organization or around any theory, like the raw food theory, are the following: (1) the existence of leaders, (2) critical thinking about the issues, (3) formation of logical rationale for opinions, (4) ability to formulate an argument and communicate it, and (5) feedback on the communication from those who agree and those who disagree. From there the feedback loop continues, perhaps in perpetuity depending on the issue and the people involved in the conflict. As long as the conflict remains healthy (meaning it is absent of backstabbing, pettiness, gossip, and a lack of candor or truthfulness), it can and should continue for as long as it takes to reach a resolution. This could mean centuries. It took physicists over 200 years to overturn many of the theories of Rene Descartes that we embraced with no exception.

Healthy Conflict in the Raw and Living Food Movement

The sections below provide an overview of some of the differences that exist within the raw and living food communities. If you have a specific interest in a topic, I have provided suggested readings and Web sites for most areas. Healthy conflicts not included here are the chocolate and sweetener controversies; they have their own chapters due to the size and complexity of the topics.

Raw versus Living

The terms raw and living are often used interchangeably, when in fact they have different definitions. Within the raw and living food movement, certain individuals support eating all or mostly living foods while others eat both living and raw foods in any proportion. Living foods are of two types. The first type is also the *ideal* living food, fresh, organic fruits and vegetables eaten in their natural state as close to the time they were picked or harvested as possible. In addition to the high nutrient value of these living foods, they also supply enzymes to help your body in digestion. You can dehydrate these foods up to about 110° F without killing the enzymes, but as we will see below, some controversy exists around this claim.

The second class of living foods, which should more accurately be called *resurrected* foods, are sprouted foods. These are seeds, nuts, and beans that have been harvested from the ground or from trees. When still in their natural state they are considered *raw*, but not living, because they are not producing enzymes. To be considered *living*, these foods need to produce enzymes. We can create this activity by soaking the food overnight (or longer) until the seeds, nuts, or beans begin generating a small sprout. Soaking causes the seeds to swell and softens the hull to allow the sprout to grow out of the seed. The sprouting action tells us that the food is now producing enzymes, so it is technically a *living* food again. We can benefit from these enzymes for about three days before they begin breaking down.

Raw food is actually a much broader category of food than living foods. Raw foods encompass all living foods plus any food in its natural uncooked state. All uncooked fruits, vegetables, seeds, nuts, and beans and uncooked natural products of these foods are considered raw foods. They continue to be raw as long as they are not heated above 110° F.

An uncooked, organic, fresh apple is both a living and a raw food. A raw almond however, is a raw food. It is not a living food. If you soak the almond overnight until it begins to sprout a little tail at the pointy end, it transforms itself into a living food. Miracles do happen. The same goes for any seeds or nuts. A lentil bean is a raw food, but it is not living unless you soak it and sprout it, at which time it becomes a living food. A raw lentil bean does not have much nutritional value because of its texture. It is impossible to eat lentil beans without them being sprouted (or cooked, of course).

Unheated agave, which is the nectar from the agave plant (a cactuslike plant) is considered raw because it is the natural by-product of a plant. However, it is not considered living because it does not contain living enzymes. And, by definition, natural by-products of plants are not considered living foods. This also includes substances like yacon syrup, raw cacao beans, nibs, or powder, carob powder, or any other food that is natural, raw, and without living enzymes buzzing around inside of it.

Juices versus Smoothies

The difference between a juice and a smoothie may seem subtle, but it is actually quite substantial. Juices result when the liquids are extracted from fresh fruits and vegetables using a juice machine. In essence, we are separating the fiber from the liquid and only drinking the liquid. Juicing of green plants in particular allows us to take optimum advantage of photosynthesis. We all remember (more or less) from grade school science that photosynthesis occurs when green plants use energy from the sun to transform water, carbon dioxide, and minerals into oxygen and organic compounds. It is one example of how people and plants are dependent on each other in sustaining life. The green pigment chlorophyll is uniquely capable of converting the active energy of light into a latent form that can be stored (in food) and used when needed. Photosynthesis provides us with most of the oxygen we need in order to breathe.[1] When we juice, all of the nutrients from the plant remain in the liquid. It's like drinking only the nutritional essence of the vegetable or fruit. As a result, we get optimum nutritional benefit with little imposition on our digestive system.

Smoothies are created by blending together whole fruits and vegetables into a thick but drinkable concoction. Smoothies retain all of the fiber *and* the liquid. Our digestive system needs to do more work, but some argue that the fiber provides our stomach with the ability to be efficient by producing adequate hydrochloric acid to allow us to maintain our acid-alkaline balance. Most smoothies contain greens. Many smoothies contain both greens and fruits, like bananas. While the fruit makes the smoothie overall more palatable, some argue that it violates the food-combining principles, which state that you should never combine fruits and vegetables. Proponents on both sides of the coin could go on forever in the juice versus smoothie debate. Most raw foodies feel strongly one way or the other. But, it's also okay to drink smoothies some days and juices on others. It really depends on your taste buds and how your body reacts to the drinks. If you are just starting to drink smoothies as part of your daily regimen, allow your body and your digestive system at least 7 to 10 days to adjust to the new experience before coming to any conclusions about whether smoothies or juices work better for you.

Salt

The first question about salt is whether to use it at all in your food. And if your answer to this is yes, then the second question is what type of salt to use. First, let's address the issue of whether to use salt or not. Raw food leaders like Brian Clement and the Hippocrates Health Institute feel strongly that salt should not be used in the living food lifestyle. They do not believe it is living and is does not add nutritional value. If you take this logic to its mainstream conclusion, salt can actually harm you. We have had the health dangers of salt drilled into our heads for decades, including hypertension, heart disease, and stroke. More importantly, when it comes to NaCl (the chemical formula for salt), it is an issue that you need to determine for yourself—again, apply your own biochemical individuality.

For most of the past 20 years, including my recent past as a raw foodie, I did not use salt on my food. Then I went to the Tree of Life for a conscious eating food-preparation course. The classes on how to prepare food did not begin until later in the week. But it didn't matter because we were kept busy with lectures from Gabriel Cousens, lectures from his colleagues, and of course, great living and raw food prepared by the apprentices and chefs at the Tree of Life. By the second day, I was so swollen, I could not wear my wedding ring and could barely tie my sneakers! First I thought it was jet lag setting in because I was not used to traveling at the time. But by the third day, when the classes began and I saw the amount of salt used in the recipes, it was *crystal* clear what was causing my swelling. First, the instructors demonstrated the correct preparation of raw vegetables. The first step in this process was massaging the vegetables with Himalayan salt. It was amazing to watch the color of the green vegetables transform to a much brighter color. And, as the instructors explained to us, the salt helped to open up the outer layer of cells, or cellulose, to make the vegetables easier for us to digest. The use of salt as an essential ingredient by the Tree of Life food preparers was stressed throughout the program. I do not believe there was one dish—even dessert, that was made without some salt. I must admit that the food at the Tree of Life was some of the best I have tasted, raw or otherwise. Given the strict principles of food preparation all of the staff follows, I am sure it is also some of the most nutritious I have ever had as well. For the large majority of the population, the use of salt would not create an issue. Here is a great example of good intention, good nutrition, and good taste banging up against a body that was saying *don't do this to me.* Now, is it possible that I could create a tolerance for more salt in my daily diet? I suppose so. But right now, this is not an option I am willing to entertain.

The arguments in favor of the use of salt in your diet include the fact that salt helps to clean the body of wastes and enhances our appetites.[2] If you have decided that salt will be part of your diet, you need to determine what type to use. Even if you only use salt sparingly, you should still use the type that is best. Regular table salt is not an option because it is reorganized and stripped of much of its trace mineral content. The salt options for the raw foodist include: sea salt, Himalayan pink salt, Celtic sea salt. Sea salt has retained its minerals. Plain sea salt is not recommended by most because it often has the same issues as table salt. So the remaining options are Celtic sea salt, a salt specially harvested off the coast of France to ensure all trace minerals are maintained, or Himalayan pink salt, mined from the mountains, and it supposedly has a richer mineral content. It also has a milder taste than most other salts. You can purchase either salt online from reputable sources and may want to try out both to see if either works for you. If you are a salter, switching to Celtic or Himalayan will provide you with greater benefits than table or regular sea salt.

Fasting (Cleansing)

As far as raw and living foodies go, fasting is where it's at—everyone agrees on that. You fast as a cure. You fast to cleanse your body. You fast for nutritional reasons. You fast to repent when even your *raw* food diet has been too acidic. You fast for spiritual reasons. You fast to increase your self-discipline. You fast to achieve mental clarity. You get the idea, right? Fasting is where it's at.

So the issue is not whether to fast, the issues are how long the fast should be and whether you should be taking more than water into your body during the fast. Dr. Joel Fuhrman's book, *Fasting and Eating for Health*, is a great overview of the medically supervised fasting process. He has been a serious faster, and has fasted on nothing but water himself for over a month, with phenomenal health benefits. Using his own experiences and his University of Pennsylvania medical degree, he has dedicated himself to making America more healthy through nutrition. You can read his book or visit his Web site for more details.

In addition to Dr. Fuhrman, others who preach the benefits of fasting include Dr. Gabriel Cousens, Norman Walker, Ann Wigmore, Brian Clement and the Hippocrates Health Institute, Natalia Rose, Brenda Cobb, Susan Schenk, and Jameth and Kim Sheridan. Some believe that water fasting for at least seven days and up to 21 days is the only way to obtain benefit from fasting. Others believe that a short fast of only two to three days provides some benefits. Depending on your body size, after either two days (for smaller bodies) or three days (for larger bodies) of fasting, your body will need to get its nutrition from its own cells. Fasting specialists like Fuhrman and Cousens believe that it is at this point that the body begins to shed old and sick cells. This is likely the greatest health benefit that fasting brings us.

The longest fasts recorded have been 40 days. Most of those were by serious fasters who were subsisting on water alone. Because fasting is usually a fasting from food, the issue is generally what liquids, besides water, should be taken during the fast? Some believe lemon water, others add green juices, some use any vegetable juices, and still others use any fruit or vegetable juice. There is also a ginger root cleanse, an olive oil cleanse, and a maple syrup cleanse. The list goes on. The best resources for understanding the nature of fasting and cleansing are Gabriel Cousens' books *Conscious Eating* and *Spiritual Nutrition* if you are ready to go the raw and living route. For a more conventional approach to fasting, read Joel Fuhrman's books or visit his Web site, www.drfuhrman.com.

Temperature Levels for Dehydrating Raw Foods

Raw food leaders appear to disagree, at the most by a range of about 12 degrees, on the maximum temperature at which to dehydrate foods. There is published documentation that 100° F is the maximum temperature to heat food to ensure enzymes live. I have also seen 105, 108, 110, and 112 as the documented maximum. I usually use the 100 to 105° F range, just to be safe.

One other recent theory that I heard about at Gabriel Cousens' Tree of Life retreat was that, to avoid fermentation, all dehydrated foods should be dehydrated for the first hour at about 135 to 140° F. Then, turn the dehydrator down to about 110° F. When I did this, my food cooked very quickly and became overly dry, so if you choose to abide by this rule, you will need to experiment with the levels and amount of time that works. Setting the initial temperature up to 140° F significantly cut down on the overall cooking time for my food, by more than half.

It is also important to note that although dehydration maintains the food in its essentially live state, there is about a 25 percent loss of energy in dehydrated food overall. Dehydration does allow you to store food for months at a time without losing overall nutritional value. Knowing the trade-offs though, you may want to decrease the overall percentage of your diet that comes from dehydrated food—if you were on this path to begin with.

Colonics

Colonics involve a cleansing of the colon of all waste materials. They have been used for years for both health and beauty purposes. Most raw foodists believe that colonics are an important part of the overall body cleansing process that goes along with eating raw food and fasting. There are a few who believe that the diet and fasting make your body so efficient that colonics are not necessary. You can find information about colonics in Natalia Rose's books *The Raw Food Detox Diet* and *Raw Food Life Force Energy*. Again, you will need to determine for yourself if you want to incorporate colonics into your regimen. The E3 Live company provides an excellent step-by-step description of the procedures and the supplies needed in their four-page laminated *permachart* entitled "Raw Foods Vegetarianism: Detoxification." It is easy to read, detailed, and highly reliable. You can purchase the permachart on their Web site for five dollars at www.E3live.com or www.permacharts.com.

Garlic and Onions

Raw foodists who oppose the use of garlic and onions (as well as any other spicy foods) do so for one of two reasons. First, they believe that the food has a disruptive effect on the digestive system and counteracts the efficiencies we are trying to instill through a raw and living diet and lifestyle. Second, they believe that these types of foods do not allow the mind to be still and therefore interfere with meditation and mindfulness practices. Those who do eat these types of foods, and there are many, do so because they like the taste, and some believe the foods have healthy medicinal qualities.

CHAPTER 10

Supplements

· ·

If you want to make your head spin, do some research to determine which supplements you need, if any. Most people walk away from a well-intentioned Internet encounter to hunt down the list of dietary supplements they need with a high level of frustration. More often than not, they may leave their search with either a multivitamin or nothing because it's just too difficult to determine what supplements will benefit them. The raw and living food movement gives a whole new meaning to supplements, often also referred to as *raw superfoods.* While it still may be a dense walk through the choices for you, there is good news. Many raw supplements or superfoods are foods themselves, not just the concentrated vitamins, minerals, and herbal supplements we are used to seeing in our local health food store. As a result, you can incorporate the supplements in your daily meal planning so it does not seem like something *extra* you are doing. The additional good-for-you nutrients can become part of your routine.

The choices can be agonizing, but after much research myself, I made some decisions on my own supplements. It may be helpful for you to understand the groundwork I did and the choices I made. Remember that everyone is unique and what works for one person may not work for another. Expect to revamp your supplement regimen at least once every 12 to 18 months. If you don't change the contents, you can stop certain supplements for a period, perhaps when fasting, just to provide a jump start to your metabolism when you begin again. You may find that your body gets bored with the same exact input week after week and month after month, so some degree of change may work better for you.

My background research on supplements included some discussions with local raw food store owners and reading recommended books. Gabriel Cousens' books and Natalia Rose's books address supplements in a helpful way. After making a preliminary list of the possible supplements I wanted to use, I attended a few lectures. The first was with David Wolfe, a nutritionist and raw entrepreneur, who has his own list of raw superfoods. Then, I attended several lectures given by Gabriel Cousens at the Tree of Life retreat where he discussed supplements and responded to many questions. My pared-down list of supplements today looks like this: three green supplements, spirulina, pure synergy, and E3 Live Blue-Green Algae flakes. I also take a Nano B (all live B vitamins) complex each day. In the winter, I take an organic vitamin D supplement and liquid vitamin C. I am currently reassessing the value of bee pollen and whether I can obtain very fresh bee pollen where I live. If I can, I will probably include bee pollen in my regimen. The beauty of this process is that as long as I plan appropriately, it's not like I am taking supplements at all. Most of these supplements are powdered or liquid, so I include them as ingredients in my morning juice, fresh salad dressing, or nut cheeses when I make them.

I have categorized the groups of supplements most commonly used by raw foodists into green supplements, bee products, minerals/vitamins/enzymes, and other superfoods. While they can only help you nutritionally, depending on your current diet, they may not be all you need to optimize your health. Most of the supplements can be purchased through various

Internet vendors. If there are limitations on where and how certain supplements can be purchased, I have noted that in the relevant section.

If you are not already on a nutritionally sound path, you may want to have a full (holistic) analysis of your blood performed to determine what deficiencies you have. The type of analysis you need is more than the typical complete blood count (CBC) that most physicians order, so you may need to visit with a physician who takes a holistic approach to healthcare or, if you are in one of the 14 states that currently licenses naturopathic doctors, visit with one of them.

Green Supplements

Green supplements are generally concentrated versions of the green and sea vegetables that are responsible for keeping our pH slightly on the alkaline side, as it needs to be for good health. If you can choose only one supplement or set of supplements, it should be green. There are many choices that I list below.

Blue-Green Algae: Also known as E3 Live algae or crystal manna, this supplement is harvested from Klamath Lake in Oregon. It is high in protein, chlorophyll, and vitamins, and it enhances both the immune system and brain function as it is naturally high in neurotransmitters. The E3 Live company, which harvests the algae from Klamath Lake and sells it, describes how the algae specifically strengthens the hypothalamus, the master gland of the endocrine system, as well as pituitary function.[1]

Chlorella: Chlorella is a high-protein algae with about 5 grams of protein in one teaspoon. It also contains high amounts of magnesium and the super detoxifier chlorophyll.[2] It contains two to five times the amount of chlorophyll as spirulina. Chlorella is the best algae for pulling heavy metals out of the system.

Spirulina: Spirulina is a complete protein and is about 60 to 65 percent amino acids. It also contains B vitamins, minerals, chlorophyll, and phytonutrients. Spirulina contains one of the highest sources of gamma-linolenic acid

(GLA)—an essential fatty acid (EFA) in the omega-6 family that is found primarily in plant-based oils). Gabriel Cousens claims that only mother's milk is higher in GLA, so he recommends that lactating mothers use increased amounts of the supplement.

Pure Synergy: Pure synergy is a powder compound of organic vegetables, fruits, and herbs. It contains 62 different ingredients including spirulina, Klamath Lake algae, chlorella, salina, kelp, Irish moss, dulse, wheatgrass, kamut, and Chinese rejuvenation herbs. It also contains several medicinal mushrooms. Dr. Cousens recommends this as a great supplement to take when traveling or camping.

Sea Vegetables (Kelp, Dulse, Nori): All sea vegetables are high in protein, amino acids, and many B vitamins. Kelp in particular is very high in iodine, which is supportive for the thyroid. Eating sea vegetables on a daily basis, in salads or just plain, is a good idea. You can purchase dried sea vegetables from most health food stores. As an additional way of getting exposure to the values of sea vegetables, and kelp in particular, I use kelp soap manufactured by the E3 Live company. It has an excellent exfoliating quality to it and by removing dead skin cells, it helps oxidize the body while you shower.

Phytoplankton: Plankton are any drifting organisms that inhabit the deep sea. Phytoplankton are the component of plankton that produce organic substances from inorganic molecules and the sun—also known as photosynthesis. Through photosynthesis, phytoplankton are responsible for 50 percent of the oxygen present in the earth's atmosphere. As a supplement, phytoplankton is bottled in a way that preserves the marine phytoplankton in suspended animation, protecting the original life energy. The high chlorophyll content of phytoplankton also increases oxygen uptake. One droplet is sufficient to get you through the day and just that one droplet tastes like the ocean floor.

Bee Products

Bee products have already been addressed in the chapters on sweeteners. There, I described both raw honey and bee pollen. In this chapter, I will not spend a lot of time describing them again, but will concentrate instead on the benefits of bee products to your health.

Raw Honey: Although not a vegan food, even David Wolfe describes raw honey in the following manner: "Honey, in its organic, wild, raw, unfiltered state is rich in minerals, antioxidants, probiotics, enzymes, and is one of the highest vibration foods on the planet."[3]

Bee Pollen: Bee pollen is said to be the most complete food found in nature. Bee pollen helps to support the immune system and is high in vitamins A, B, C, and E, minerals, proteins, amino acids, hormones, and enzymes. Bee pollen has also been shown to have a protective effect against radiation. Gram for gram, pollen contains an estimated five to seven times more protein than meat, eggs, or cheese. In addition, the protein in pollen is in a predigested form and is easy to assimilate into the digestive system.[4]

Vitamins/Minerals/Enzymes

Crystal Energy: Crystal Energy, a detoxifier, is a patented form of minerals, rich in silica. The manufacturer of the substance claims that Crystal Energy helps remove heavy metals from the body. They also state that it challenges the symptoms of dehydration and minimizes the process of aging. Gabriel Cousens includes Crystal Energy on his list of supplements to take when you will be traveling by plane. He believes it helps keep the body in balance in compressed airplane cabins as well as to help fight off the effects of jet lag. You can read more about the details of how Crystal Energy was developed by Patrick Flanagan on his Web site, www.wetterwater.net.

MSM: In *The Miracle of MSM: The Natural Solution for Pain,* the authors Dr. Stanley W. Jacob, Dr. Ronald M. Lawrence, and Martin Zucker claim MSM relieves back pain, headaches, muscle pain, arthritis, allergies, fibromyalgia, and more. MSM is a sulfur-based, naturally occurring compound with its roots in oceanic phytoplankton. The closest chemical relative to MSM is dimethyl sulfoxide (DMSO). MSM is available as a capsule or crystal to be taken orally. In addition, it is available in cream or lotion. Although I do not take MSM as a supplement, I use organic MSM-based shampoo, conditioner, and face wash.

Nano B Complex: This is a liquid that contains living, not synthetic, B vitamins to support the liver, immune system, heart, brain, and mood balance. The substance contains vitamin B_1, B_2, B_3, B_5, B_6, B_{12}, folic acid, para-amino acid, and biotin. This product is made by Premier Research Labs. Individuals on a vegetarian or vegan diet cannot obtain their B vitamins from food sources since they occur in meats or animal-based foods. Although the body has some ability to manufacture vitamin B_{12} and other B vitamins naturally, not everyone's body can assimilate these B vitamins well (I happen to be one of them), so the Nano B complex can help give vegans the advantage they need in this area. This is why I incorporate it into my regimen.

D_3: Vitamin D is best obtained directly from sunshine. However, in certain climates during the winter months, this is not possible. Vitamin D supplement may be necessary during those times. A natural, organic (nonsynthetic) version of vitamin D is the best. Dr. Michael F. Holick, professor of medicine, physiology, and biophysics at Boston University Medical Center and author of *The UV Advantage: The Medical Breakthrough that Shows How to Harness the Power of the Sun for Your Health,* recommends at least 1,000 IU of vitamin D daily. He recommends the best way to obtain vitamin D during winter months or cloudy times is by getting exposure at a tanning salon that uses low-pressure lamps. The specific request for low-pressure lamps is important. Dr. Holick claims that you only need about 25 percent 1MED exposure in a tanning booth to obtain 1,000 IU of D. He defines 1MED as the amount of time it would take you to get *pink* in the sun. So, if it takes you 30 minutes to get pink on average when you are in the sun, and you need 25 percent of that time in a tanning booth to get your daily requirement of vitamin D, you will spend seven and a half minutes a day in a tanning booth.[5]

NCD (also called Zeolite): NCD stands for natural cellular defense. It is an energy substance made from a compound that pulls out pesticides, radiation, metals, and other chemicals from the body. It has been shown to have a protective effect against diseases like hepatitis C. Zeolites are extracted from natural volcanic rocks. They have a crystalline shaped like a honeycomb that helps trap and remove the heavy toxins and metals from the cells in your body in a safe, natural way. It is one of the few minerals that is negatively charged. As a result, zeolites apparently attract and draw the harmful metals and toxins in your body to it, capture them, and help your body remove them.

Vitalzym X: Vitalzym X helps boost enzyme production and is good for inflammation and injury. It boosts the immune system and has an anticancer effect. Vitalzym X is a blend of enzymes and nutrients formulated for optimal energy. It contains these enzymes to aid in digestion: proteinase, bromelain serrapeptase, amylase, and lipase. The supplement helps maintain normal enzyme levels and it helps balance your body's own repair mechanisms. There is a less powerful version of the supplement called simply *Vitalzym*. Vitalzym can be purchased directly over the Internet. However, Vitalzym X requires a prescription from your healthcare provider due to its potency.

Other Superfoods

Camu-Camu: Camu-Camu (Myrciaria dubia) is a native bush to the South American rainforest. It is higher in minerals than any other Amazonian plant. A fresh berry, it contains 30 to 60 times more vitamin C than an orange. The powder is typically found in vitamin C supplement capsules. The Camu-Camu berry is a source of phosphorus, calcium, potassium, iron, and the amino acids serine, valine, and leucine. Small amounts of the vitamins thiamine, riboflavin, and niacin are also present.

Maca: Maca was also addressed in chapter 6 on sweeteners. The organic maca root has been used in Peru since before the times of the ancient Incas. The maca plant is similar to a radish or turnip and it can have different hues of color on its skin including red, purple, yellow, or cream and black. Maca is a protein-rich food. It also has 18 amino acids, vitamins (A, B_1-thiamine, B_2-riboflavin, B_3, niacin, B_6, B_{12}, C, E, carotene), and minerals that include calcium, phosphorus, iron, zinc, magnesium, copper, sodium, potassium, selenium, boron, manganese, and aluminium.

Probiotics: Probiotics are dietary supplements containing beneficial bacteria. According to the World Health Organization, probiotics are live microorganisms that when administered in adequate amounts confer a health benefit on the host.[6] Probiotics grow naturally in fermented foods like sauerkraut and kimchi. Claims are made that probiotics strengthen the immune system to combat allergies, excessive alcohol intake, stress, exposure to toxic substances, and other diseases. Powdered probiotics labeled as *a natural whole food,* with no additives or flavoring are the best to use. You can simply sprinkle a little powder into a drink or recipe once a day or a few times a week to receive a benefit. You can purchase probiotics in the refrigerated supplement section of Whole Foods and most health food stores.

Conclusion

This chapter is not meant to be a complete listing of supplements—raw or nonraw. Rather, it is a compilation of those supplements recommended by at least three of the raw food leaders. It is best to try one supplement at a time to determine if it is beneficial to you. This may take weeks before you have enough data to determine whether you feel better or are more energetic or efficient as a result of a supplement. After you have determined the usefulness of the first supplement, it is best to add supplements one at a time.[7]

CHAPTER 11

Basic Raw Food
Preparation 101

· ·

This chapter is to the raw foodist as boiling water is to a chef who prepares cooked food. It's a basic introduction to the tools, the supplies, and the mind-set you need to prepare raw foods. There are many excellent raw food preparation guides or *un-cookbooks* that have been published. I refer you to these books by way of appendix A with a list of resources I have personally purchased, read, and used. The authors of many of these books, like Jennifer Cornbleet, Matt Amsden, Jeremy Saffron, Renee Loux Underkoffler, Terces Englehart, the Jubbs, Alissa Cohen, and Sarma Melngailis, have included additional details about supplies and methods that you will find helpful. In keeping with the theme of this book, it is my intent to provide you with good, solid, reliable introductory information to every aspect of the raw and living food philosophy. You can research the parts you find more interesting and, most importantly, decide for yourself which of the concepts, if any, you want to incorporate into your own life.

Your Mind-set

The first thing you need to understand about raw food preparation is that it can be as simple or as complex as you want it to be. Each raw food chef has her own approach to the food preparation process. Raw chefs like Sarma Melngailis and Renee Loux Underkoffler, both who own or have owned raw restaurants, take a more haute cuisine approach to their craft. Their food is great, but you will need to plan ahead in both time and specific purchases if you are going to tackle a few of their recipes. Folks on the other end of the spectrum include Jennifer Cornbleet, whose book, *Raw Food Made Easy*, is both simple and wholesome, as well as Carol Alt, a new raw food convert. Alt's conversion and real-world need to make the lifestyle work for her caused her to find some of the best and easiest recipes. Finally, there are the raw chefs who fall in the middle. They tend to draw from both the haute cuisine style as well as the eating-over-the-counter crew. These raw chefs include Matt Amsden and his *RAWvolution: Gourmet Living Cuisine* book, Jeremy Saffron and his *Raw Truth* book and Lori Baird and Julie Rodwell's *The Complete Book of Raw Food*. Most of these folks have Web sites where you can test out some of their recipes. You won't need to purchase every book, although once you have decided on your favorite raw chef, you will probably want to invest in the chef's un-cookbook.

After testing some books and recipes and becoming more familiar with and confident in your food preparation skills, you may want to develop your own style. Because raw food preparation is this seemingly new concept, many of us feel like we need a book or recipe to tell us how to do everything. In fact, the opposite is true. Think about your current or preraw days. Did you need someone to tell you how to make a grilled cheese sandwich or grill a hamburger? Once you feel at home with all of the *raw food materials*, you should be able to create the majority of your meals on your own. I like to use Matt Amsden's raw mashed potatoes as an example. The concept of using raw macadamia nuts, fresh cauliflower, garlic, and olive oil to create a raw version of mashed potatoes astonished me. But I tasted them and they were delicious! I followed the raw mashed potato recipe from the *RAWvolution* book twice.

Eventually, I made some modifications in the recipe. Some of them were terrible, but you need to be willing to take that risk to make raw food preparation your own. While I usually use the original recipe, I sometimes use almond milk or avocado oil for a softer flavor. The possibilities are endless.

Feeling at Ease with Your Food Preparation Responsibilities

I have always been pretty skilled at eating food. But the preparation part was a whole different story for me. I consider myself to be a recovering perfectionist, but for decades the task of preparing a meal or dessert was quite daunting to me. What if I did not have just the right ingredients? What if I measured something wrong? What if the oven heated unevenly? And more importantly, what if the meal was a flop? My trip to the Tree of Life Conscious Eating program changed all of that. During one of the food preparation classes, I asked a question about exactly how many cups of sprouted almonds should be used in making fresh almond milk. The answer I received was not what I expected to hear.

The instructor said a few things. First, he said, you need to consider what you will be using your almond milk for—drinking, cereal, mashed potatoes (they use a different recipe than Matt Amsden), ice cream, etc. Based on the use, you may want your milk to be of a certain consistency—thick and creamy for ice cream, thin for drinking or cereal. Then the instructor said, "I can give you general guidelines, but you really need to figure out what works best for you." Then he talked about certain proportions of nuts to water that could be used, but stressed that there were no hard and fast rules. Third, and most important, as you are making the milk, test very small quantities throughout the process so you can modify your ingredients if necessary. I still wanted to know where exactly I would find the information to tell me when a modification was necessary. As the week came to an end, I did become more comfortable with all of these instructions. Raw food preparation is not complicated. In fact, most of it is about simple activities that come naturally to all of us: cleaning and cutting vegetables and herbs; combining lettuces and pumpkins or sunflower seeds to make a salad a complete food;

and blending carrots, avocado, and sweet potatoes for a hearty soup. Most of the other activities, save the cleaving of the young Thai coconut, we can learn easily. These include how to operate a blender, food processor, or dehydrator. Making and eating the most nutritious food possible should create a pretty positive experience for you.

Coming to Terms with the Young Thai Coconut

The young Thai coconut is a paradoxical, blissful food. By blissful, I mean it brings bliss to all who eat it, especially the coconut pulp and the food in which the pulp is an ingredient, such as raw ice cream, raw whipped cream, and coconut noodles. By paradoxical, I mean that while it is the primary ingredient in some of the most indulgent and best-tasting foods, raw or otherwise, it is also the most difficult to obtain (in its natural state) and the most difficult to deconstruct for cooking purposes. Some of the many divine foods I have tasted that are made with young Thai coconut pulp as the primary ingredient include raw ice cream, pumpkin pie, raw whipped cream, and raw noodles. Many of the raw food preparation books describe the process of opening and cleaning out the coconut. Sarma's book *Raw Food Real World* provides a description and some accompanying pictures.

While at the Tree of Life, I had the opportunity to observe some food apprentices preparing their graduation meal for the following day. It was evening and besides me and one other Internet user, they were the only ones in the café (the only location at the Tree where you can pick up an Internet signal). While I worked, I watched these three newbie raw chefs cleave, open, and clean about 70 young Thai coconuts that would be used as the base for tomorrow's lunch dessert—raw vanilla ice cream. The process took two strong young men and two big scary cleavers to open the coconuts and one female chef to scoop out the pulp. It took four hours and three people to partially prepare ice cream for about 40 people. The economics of that just don't compute. However, the ice cream was phenomenal and put a major positive spin on everyone's mind-set the next day. So the outcome was a success.

Giving Thanks and Taking Time while You Eat

Eating over the counter, and sometimes even over the sink, has been a fairly common habit in my house over the past two decades. This practice is the result of conflicting schedules, fast food, doing more and more (for what?), and probably just plain old habit. Lately, my family and I have been using the table to sit down and eat, and we hope to continue this practice. Most raw food programs I have attended include a short prayer or statement of *thanks for the abundance* before the meal. It doesn't take a lot of time, and it does not need to be religious, but giving thanks and taking time to eat helps you to be more conscious in your approach to your food and eating. This consciousness begins to naturally flow into other areas as well.

Eating Mindfully: Really Tasting the Food and Experiencing Eating

Being mindful is the natural consequence of being conscious. In some ways, you can see mindfulness as the next step after you have become conscious of your food preparation and eating habits. Jon Kabat-Zinn from the University of Massachusetts Medical Center founded the hospital's Center for Mindfulness. He has written numerous books on the topic of mindfulness, including *Wherever You Go, There You Are, Full Catastrophe Living,* and *Mindfulness Meditations.* In his books, Jon Kabat-Zinn describes one of the first activities included in his mindfulness and stress-reduction workshop.

Participants are asked to eat one raisin *mindfully.* This process takes the group about eight minutes. Yes, eight minutes to eat one raisin! They examine the raisin first, describe how it looks and feels. They smell it. They are aware of every movement as they put it from their hand into their mouth and begin to chew. They count the number of chews (a practice also common in the macrobiotic diet). One of the interesting things that Dr. Kabat-Zinn does is ask how this activity differs from what these folks would normally do with raisins. The group usually responds by cupping one of their hands and explaining that ordinarily they would have scooped and shoved a handful of raisins into their mouths. Before they even swallowed the raisins, they would have another handful ready. Dr. Kabat-Zinn then asks (rhetorically) 'What's going on here?"

It's easy to appreciate a raw food meal for many reasons. First, it is healthful. Second, it benefits many others by using organically grown and, hopefully, locally grown local foods. And third, it is colorful and beautiful. Most raw food meals contain shades of bright green, yellow, red, and orange. This can make anyone smile. And, just think about what this meal looks like compared to a meal of French fries and a hamburger from a fast-food drive-through. But herein lies the issue. In order to appreciate the differences, you must first *think* about it. This may be something you choose to do, or not.

Your Tools

Most raw food recipe books contain a complete listing of all of the tools you need to start your raw kitchen. With the exception of a juicer, you may already have many or most of these tools. I recommend *Raw Food Real World* and *The Complete Book of Raw Food*. Both of these books have pictures, which makes it easier the first time around. Below is an overview of the tools you will need and some helpful hints I have picked up along the way. For more complete information on raw food tools, please reference one of the books I have noted or their Web sites. Appendix B includes a list of all of the good, reliable Web sites I use to obtain raw food supplies, tools, clothing, and snacks.

Small tools. Your primary small tools will be your knives. You need a few different kinds because you will be cutting and chopping a lot of stuff. First, if you can splurge on a knife, I recommend a ceramic knife. They are highly breakable and cannot be used to cut hard veggies like beets and carrots. But they are great for all green vegetables, most fruits, and all fresh herbs. The ceramic is particularly important because stainless steel or metal can oxidize vegetables, and ceramic will not do this. If you will be working with the young Thai coconuts, you will need a strong cleaver (and strong arms). Other small tools you may need are bowls and storage bins for your dehydrated foods and for storing your nuts and seeds.

Large tools or small machines. Tools in this category include a spiral slicer (used to slice zucchini to use as spaghetti, a big raw favorite), a mandolin (used to slice zucchini and other squash for use as flat noodles), a coffee grinder (to chop flax and other seeds), and, if you are planning to make any fermented foods—all which are very high in probiotics like kimchi or raw sauerkraut—you will need a ceramic fermentation pot.

Machines. You will need a good high-speed blender (I recommend a Vita-Mix blender—it is so strong, it can blend a wooden brick), a food processor, a juicer (I use a Champion), and one or more dehydrators, depending on how many people you are feeding. The Excalibur is the best dehydrator to use for raw food *cooking*. Dehydrators are sold by the number of trays contained in the dehydrator box. You can purchase a four-, six-, nine-, or even a 12-tray dehydrator. The trays are stacked fairly close together in the dehydrator box, so if you are planning to dehydrate thicker foods like vegetable patties, biscotti, or even raw bread, you will only be able to use every other tray in the dehydrator. If you buy a six-tray dehydrator, you may only have three trays available for use. Please see appendix B for a list of online stores that sell all raw food supplies for more detail. The raw food cookbooks include helpful resource information also. These machines can be quite expensive, so you may decide to purchase a juicer or a blender, but not both initially.

Your Raw Food Supplies

In addition to fresh vegetables and fruits, you will need certain raw food supplies to get started with the raw food preparation process. Key among these are nuts and seeds. For those transitioning from a nonraw to a raw diet, nuts and seeds play an important part in providing protein. When ground and dehydrated, they can also be used as meat (in the form of veggie patties) or bread (in the form of biscotti or flatbreads) substitutes. Particularly in the cold months, nuts and seeds provide a fullness that you don't get from fruits and vegetables. You can buy organic nuts from your local Whole Foods store. If you are looking to buy nuts in larger quantities with wholesale discounts, visit www.sunorganicfarm.com They carry all nuts, herbs, seeds, and other supplies. Because they carry both raw and nonraw foods, you will need to make sure you select the raw version of the product you are ordering. The type of nuts and seeds you buy depends on your taste preferences. I usually keep almonds and macadamia nuts (for me) and cashews (for my husband) at all times. Good choices in seeds include: sunflower seeds (you can use them in many of the sauce and seed-cheese recipes and you can sprinkle them on salads), golden flaxseeds (a great source of omegas, you just need to make sure you grind them before eating; the body cannot process flaxseed whole), black sesame seeds (I often add them to vegetable recipes), and chia seeds (they are full of nutrients and make a great tapioca-like pudding when mixed with almond milk and stevia or agave).

The last category is herbs. You can and should use as many and as much *fresh* herbs as possible. Not only do they taste great, they are also highly nutritious when they are fresh as opposed to dry. You will also need to use about 10 times more fresh herbs than if you were using the dried version. My favorite is fresh cilantro. It is loaded with vitamins and is a great brain-support herb. I have spoken to people with multiple sclerosis who swear that drinking fresh cilantro juice has improved their symptoms tremendously. One woman at Arnold's Way, who treats her MS with fresh cilantro juice, was able to go from a wheelchair to a walker. She credits the cilantro.

THE *lifestyle*

· · · · · · · · · · · · · · · ·

CHAPTER 12

Health and Healthcare in the Raw and Living Food Lifestyle

A natural by-product of the raw and living food lifestyle is increased health. This translates to decreased sickness, which in turn translates to decreased medical and healthcare intervention. While we all know that eating food that is good for us will result in better health outcomes, we still seem to need more evidence or some other motivation to take action. This mind-set may have something to do with the instant or quick results we are promised by more than 100 *diets* on the market today. Eating well and practicing a balanced lifestyle, like that supported by the raw and living food movement, does not produce results overnight. It takes some time. It takes some discipline. Mostly, it takes the guts to do the thing that is right for you and for those that you love. It means we have to ignore 80 percent of the aisles in the grocery store, drive by every fast-food restaurant we see, and if we choose the path, eat our food in its fresh, natural, uncooked state.

Many people in their last stages of cancer turn to raw food as a cure. The Hippocrates Health Institute and the Ann Wigmore Foundation, both founded by Ann Wigmore, were created to use fresh foods in their natural, uncooked (living) state as a way of rebuilding health—physically, mentally, and emotionally.[1] These organizations, as well as their spawn, attract tens of thousands of people per year from around the world, most coming to heal themselves of a long-term chronic condition or a newly diagnosed acute condition such as cancer. The total number of individuals that these organizations have helped in the 20 years since their original founding is impossible to state for sure. But what we do know is that there have been many documented cases. Even Brenda Cobb, who runs the relatively new Living Food Institute in Atlanta, Georgia, has amassed over 40 testimonials from students who have attended her living and raw food workshops in the past year. One student calls the experience *transformative regeneration.* All experienced some improvement in their health as a result of learning to prepare raw foods and then incorporating them into their diets.

Perhaps even more curious is that these treatments are not paid for by health insurance. The living food lifestyle, as originally proposed by Ann Wigmore and Norman Walker, is premised upon each of us taking personal responsibility for our own health and healthcare. The case studies of organizations like Hippocrates and the Wigmore Foundation are examples of individuals who have taken charge of their own healthcare. Like Wigmore and Walker, we all need to feel and believe that we are personally responsible for our health and that our decisions impact the outcomes we get.

While there is not a specific *healthcare* program for the raw and living food lifestyle, the following are consistent with the lifestyle. Taken together, one might hypothesize that these four criteria represent the health principles of the raw and living philosophy.

1. Focus on prevention and healing.
2. Take a natural approach.
3. Identify the cause of any symptom or problem and work to eliminate it.
4. Use nutritional solutions whenever possible.

In this chapter, we discuss some of the roots and contributions of the raw and living food lifestyle and its health philosophy. We will also look at individuals who have made significant contributions to advancing the raw and living food lifestyle by reaching out to people who are newly diagnosed with an acute illness or who are suffering from some chronic health issue. We'll look first at the individual physicians and organizations that have used raw foods to heal. Then we will look at healthcare practices that are consistent with the raw and living food lifestyle, and finally, we will profile organizations' successes with healing through raw food. Where these organizations have been willing to share their data on their impact on others' health, we will share that data with you as well.

Physicians and Clinicians Who Set the Stage for a Raw Lifestyle System of Healing

There is a group of clinical pioneers who I consider to be a part of the nutrition-based healthcare movement. The commonality among these pioneers is their belief that there is a direct, causal link between diet and health. Otto Warburg, Max Gerson, and Ann Wigmore all started their careers with a focus on how to use diet to cure cancer. Most members of the group extended the causal dietary link to include other maladies. All also believed that, once diagnosed, the body must be *detoxified* and then eased back into a diet consisting of only fresh vegetable juices and fresh, uncooked food in its natural state. Today we would add the adjective *organic,* but for most of these pioneers, the issues we face today with chemicals and pesticides were not concerns for them.

Each one of these nutritional healthcare pioneers created significant, reliable documentation about their work and any studies they conducted. Most of this documentation was in the form of books or manuscripts that have been published since their death. Some of them, like Otto Warburg, who won the Nobel Prize, were acknowledged worldwide for their accomplishments. Others, like Max Gerson, just kept plugging along with patient success after success, documenting the cases and his methods along the way. Today we are fortunate to have physicians like Gabriel Cousens, educated in mainstream America as well as through legitimate holistic and spiritual institutions, leading the way to help define healthcare principles of the raw and living food lifestyle. Dr. Cousens has documented all of his theories in comprehensive books. Most recently, he has conducted an experiment, using scientific methods, to show the impact of the raw and living diet on patients with diabetes. The preliminary results of this study are published in his book, *There Is a Cure for Diabetes: The Tree of Life 21 Day+ Program.*

Otto Warburg, MD, PhD (1883-1970)

As described in chapter 8 on pH balance, Otto Warburg won the Nobel Prize in 1931 for his findings that cancer cells thrive in oxygen-deprived conditions, or when the body is acidic. Cancer, he said, is a defect in how a body breaks down glucose. He also said that "The prime cause of cancer is the replacement of the respiration of oxygen (oxidation of sugar) in normal body cells by fermentation of sugar."[2] Until the 1960s, Warburg was still lecturing on the relationship between diet and cancer. The components of his theory on cancer are: (1) avoid acidic (low oxygen) foods—or, in the alternative, eat more alkaline foods, (2) take in plenty of oxygen and fresh, clean air, and (3) don't eat sugar or anything containing sugar.

The Nobel Committee expected the world to benefit from Warburg's vital discoveries about cancer. However, cancer was not eradicated. Some say that Warburg's big mistake was describing the prevention of cancer in too simple a format. Rather than describing his work on a theoretical level, Warburg described the exact conditions in the cells that set up and cause cancer. By sharing the practical application, Warburg made it possible for others to develop functional, practical ways to inhibit the development of cancer. Many raw and living food organizations, such as the Hippocrates Institute and the Ann Wigmore Foundation, have done just this. Despite opinions disputing the validity of Warburg's work, no one has ever disproved the validity, correctness, or applicability of his discoveries.

Brian Peskin's research, published in 2007, was based upon Warburg's principles, reviving the importance of Warburg's work. Through Peskin's work, the work of raw and living health programs like Hippocrates Health Institute, and the recent publication of *The Hidden Story of Cancer,* a book based on Warburg's findings, perhaps a larger segment of the population will at last benefit from Warburg's award-winning work.

Max Gerson, MD (1881–1959)

Right around the time that Otto Warburg was receiving the Nobel Prize for his discoveries linking diet to cancer, Max Gerson, also from Germany, was linking diet to migraines and tuberculosis. Max Gerson, who himself suffered from severe migraines, discovered that a change in diet prevented the onset of these crippling headaches. He began seriously pursuing diet as a way to heal disease when he placed one of his patients on the migraine diet and the patient was cured of skin tuberculosis. His therapy became well-known as a cure for tuberculosis. He also successfully treated Albert Schweitzer's adult onset diabetes through diet. In 1938, Gerson immigrated to the United States and treated hundreds of cancer patients who had been given up to die after all conventional treatments had failed. In 1958, based upon 30 years of experiments, Dr. Gerson published his theory, treatment plan, and case studies in his book, still available today, *A Cancer Therapy: Results of 50 Cases.* Gerson treated all of his patients using the following four principles:

1. Diet has a considerable effect on almost all diseases.
2. The human body can heal itself if given the appropriate nutrients.
3. Any effective treatments for degenerative disease must treat the whole person.
4. People with serious illnesses must help their body detoxify.

Although he does not use the term *raw diet,* Dr. Gerson's cancer diet includes the following components: (1) fresh fruit and vegetable juices all day long (he recommends using the Champion juicer), (2) the diet is fruit and vegetable based, with cooking only over very low heat without using water (when cooked at all), (3) apples should be eaten in every form—raw, cooked, grated, juiced, etc. In 2007, Gerson's daughter, Charlotte Gerson, published *Healing the Gerson Way: Defeating Cancer and Other Chronic Diseases.* The Gerson approach is a holistic one that considers the impact of stress, body-mind connection, and the environment on our health. One chapter in Gerson's book, entitled "Happy Foods," includes foods from the *vegetable kingdom,* and states, "Besides being lighter, more pure and easier to digest and absorb, each one contains a subtle mixture of vitamins, enzymes, minerals and trace elements, which work in synergy and supply the depleted organism with valuable nutrients."[3]

Ann Wigmore, ND, DD (1909–1994)

Dr. Wigmore is the pioneer of a raw diet based on wheatgrass and sprouts that took hold in the 1960s and continues to this day through three different institutes in New Mexico, Florida, and Puerto Rico despite her death in 1994. Wigmore's philosophy of healthcare is apparent in the title of one of her books, *Be Your Own Doctor.* In this book, she includes a listing of 21 testimonials from patients who were healed using the natural diet-based methods that she supports. The conditions that these patients suffered from include arthritis, asthma, constipation, diabetes, emphysema, leukemia, and Parkinson's disease.

Wigmore's focus is on becoming self-sufficient and not being dependent on others for health treatment or foods. She instructs the reader in methods about how to grow and prepare one's own food. She also provides opinions on activities in addition to diet. These activities include interacting with the environment, keeping a good posture, engaging in regular exercise, relaxing, and sleeping well. She also believes that we can obtain significant benefit from massages, acupuncture, music, meditation, and solitude. Her insights are helpful in giving a holistic, balanced perspective about healthcare.

Norman Walker, ND (1886–1985)

Dr. Walker's approach to healthcare was also diet-based with a focus on juicing, a process that he invented. His book, *Vibrant Health,* is an excellent resource, as it walks the average person through the basics of human anatomy and physiology. An individual with no scientific background can benefit from the basic information he includes in the book. Walker takes a holistic approach by including a chapter on mental health. In the book, he summarizes the Dr. Walker program as following these seven principles:[4]

1. Drink raw vegetables and fruit juices.
2. Keep the colon clean.
3. Drink pure water.
4. Exercise regularly.
5. Eliminate unhealthy products (alcohol, tobacco, etc.).
6. Control your weight.
7. Take care of your spiritual and emotional health.

Gabriel Cousens, MD, DD, MD(H)

Of all of Dr. Cousens' nutrition-based healthcare contributions, his book *There is a Cure for Diabetes* is the most notable. In this book, Cousens follows the cases of 11 patients with diabetes who were placed on the Tree of Life diet. Patient data regarding initial, ending, and point-drop fasting blood sugars were collected and analyzed. The overall hypothesis of the study was whether there is a positive relationship between the Tree of Life diet and the amount of insulin the patients need to take. In other words, did insulin requirements decrease as a result of the diet? Every patient had a lower ending fasting blood sugar than the beginning fasting blood sugar. All patients showed some improvements and all had some decrease in their insulin intake. While the number of patients in the study is insufficient to support a scientifically valid conclusion, it does provide a sound start for data collection on this subject. Dr. Cousens continues to work with patients through his program at the Tree of Life. It is very likely Dr. Cousens will be the first clinician-scientist to obtain widespread support for a diet-based therapy to treat a chronic condition.

T. Colin Campbell, PhD

Although Dr. Campbell is not a raw food practitioner, he is a vegan. Most importantly, he has led the first study using scientific methods to show a relationship between diet and health. Using initial data from a cancer study conducted in China in the 1970s (at the time Premier Chou En-Lai was dying of cancer), Campbell and his team conducted a comprehensive study of the relationship between diet and cancer. Over 6,500 adults were enrolled in the study, and tests were conducted on them before and after they ate. When Dr. Campbell's team finished analyzing the data, they had over 8,000 statistically significant associations between lifestyle, diet, and disease. A read of the entire book, *The China Study*, is necessary to get a feel for the report. However, there are eight principles of food and health that came out of the study. They are:

1. Nutrition represents the combined activities of countless food substances. The whole is greater than the sum of its parts.
2. Vitamin supplements are not a panacea for good health.
3. There are virtually no nutrients in animal-based foods that are not better provided by plants.
4. Genes do not determine disease on their own. Genes function only by being activated, or expressed, and nutrition plays a critical role in determining which genes, good and bad, are expressed.
5. Nutrition can substantially control the adverse effects of noxious chemicals.
6. The same nutrition that prevents disease in its early stages can also halt or reverse disease in its later stages.
7. Nutrition that is truly beneficial for one chronic disease will support health across the board.
8. Good nutrition creates health in all areas of our existence. All parts are interconnected.[5]

Complementary and Alternative Medicine (CAM)

Complementary and alternative medicine (CAM) is a group of diverse medical and healthcare systems that are not presently considered conventional medicine in the United States. The National Center for Complementary and Alternative Medicine (NCCAM) is the federal government's lead agency for scientific research on CAM. Their mission is to explore complementary and alternative healing practices in the context of rigorous science, train CAM researchers, and disseminate authoritative information to the public and professionals. NCCAM's Web site (www.nccam.nih.gov) contains many extensive resources if you are interested in exploring CAM further.

Ayurvedic (I-er-vay-dik) medicine, a CAM that has its roots in India and Hinduism, is over 2,000 years old. Ayurvedic medicine principles are consistent with the raw and living foods approach to healing and healthcare. While it is common for us in the West to want to categorize and separate things into neat packages, Eastern medicine (which predominates much of CAM) is more about free flow and balance. The names given to various healing techniques do not preclude them being used with healing techniques that are grouped in another category.

Ayurvedic Medicine

Ayurvedic medicine integrates the body, mind, and spirit to prevent and treat disease and to promote wellness and is one of the world's oldest medical systems.[6] In Ayurvedic philosophy, people, their health, and the universe are all thought to be related. Health problems can result when these relationships are out of balance. The goal of Ayurvedic medicine is to cleanse the body of substances that can cause disease and to re-establish harmony and balance. According to a NCCAM survey, about 751,000 people in the United States have tried Ayurveda, and 154,000 people had used it within the 12 months preceding the survey.

Ideas about relationships among people, their health, and the universe form the basis for how Ayurvedic practitioners think about problems that affect health. Ayurveda holds that: All things in the universe (both living and nonliving) are joined together. Every human being contains elements that can be found in the universe. All people are born in a state of balance within themselves and in relation to the universe. This state of balance is disrupted by the processes of life. Disruptions can be physical, emotional, spiritual, or a combination. Imbalances weaken the body and make the person susceptible to disease. Health will be good if one's interaction with the immediate environment is effective and wholesome. Similar to quantum theory, Ayurveda medicine believes that disease arises when a person is out of harmony with the universe.

Food Principles

Food is a common theme throughout Ayurvedic medicine. All food is divided into six types so that balance can be used in food preparation and eating by using foods from the different categories. Ayurveda also categorizes everyone into a *dosha*, or personality and body type. Based on your *dosha* (addressed later), certain foods are more balancing for you than others. The six categories of food include:

1. Salty (celery, sea vegetables, olives, adzuki miso)
2. Acid (kimchi, fermented foods, grapefruit, tomatoes)
3. Fat (avocados, coconut, nuts, seeds, oils)
4. Sweet (stevia, fruit, beets, carrots, squash, mesquite, carob)
5. Spicy (peppers, tamari, garlic, onion)
6. Bitter (cacao, tumeric, dandelion)

Doshas

Three qualities called *doshas* form important characteristics of the individual's constitution and body type. Practitioners of Ayurveda call the *doshas* by their original Sanskrit names: *vata, pitta,* and *kapha.* Each *dosha* is made up of one or two of the five basic elements: space, air, fire, water, and earth. Each *dosha* has a particular relationship to body functions and can be upset for different reasons. A person has her own balance of the three *doshas*, although one *dosha* usually is prominent. *Doshas* are constantly being formed and reformed by food, activity, and bodily processes. If you are interested in determining your dominant *dosha*, John Douillard provides an excellent and simple body type questionnaire in his book, *The 3-Season Diet: Eat the Way Nature Intended*.

- The *vata dosha* is a combination of the elements space and air. It is considered the most powerful *dosha* because it controls basic body processes such as cell division, the heart, breathing, and the mind. *Vata* can be thrown out of balance by staying up late at night or eating before the previous meal is digested. People with *vata* as their main *dosha* are thought to be especially susceptible to skin, neurological, and mental diseases. Douillard refers to this *dosha* as *winter* type—a type who has an aversion to cold weather and light, has a thinner build, and has tendencies toward dry skin, worry, constipation, and interrupted sleep.[7]

  Examples of foods that are more likely to balance the *vata dosha* include dried fruits, apples, cranberries, pears, broccoli, cabbage, cauliflower, onions, peppers, barley, buckwheat, millet, and oats. Examples of foods that may create imbalance in the *vata dosha* include avocado, bananas, melons, pineapple, carrots, green beans, radishes, rice, wheat, most meats, and seafood.

- The *pitta dosha* represents the elements fire and water. *Pitta* controls hormones and the digestive system. When *pitta* is out of balance, a person may experience negative emotions (such as hostility and jealousy) and have physical symptoms (such as heartburn within two or three hours of eating). *Pitta* is upset by, for example, eating spicy or sour food; being angry, tired, or fearful; or spending too much time in the sun. People with a predominantly *pitta* constitution are thought to be susceptible to heart disease and arthritis. Douillard refers to this *dosha* or body type as *summer* type—a type with an aversion to hot weather, prefers cold food and drinks, can't skip meals, is a good public speaker, and has an enterprising and sharp character.

Examples of foods that are more likely to balance the *pitta dosha* include grapes, lemons, papaya, carrots, garlic, onions, radishes, spinach, buckwheat, brown rice, rye, beef, and egg yolks (if not vegan). Examples of foods that may create imbalance in the *pitta dosha* include apples, coconut, pineapple, raisins, brussel sprouts, cabbage, celery, mushrooms, peas, barley, basmati rice, wheat, chicken, and egg whites (if not vegan).

- The *kapha dosha* combines the elements water and earth. *Kaphas* are strong and have tough immune systems. An imbalance in the *kapha dosha* may cause nausea immediately after eating. *Kapha* is aggravated by, for example, sleeping during the daytime, eating too many sweet foods, eating after one is full, and eating and drinking foods and beverages with too much salt and water (especially in the springtime). Those with a predominant *kapha dosha* are thought to be vulnerable to diabetes, gallbladder problems, stomach ulcers, and respiratory illnesses such as asthma. Douillard refers to this *dosha* as *spring* type—a type with a solid, heavier build, greater strength, and endurance and is slow and methodical with a tranquil, steady personality.

Examples of foods that are likely to balance the *kapha dosha* include avocados, bananas, melons, pineapples, plums, cucumbers, sweet potatoes, tomatoes, zucchini, oats, wheat, beef, and seafood (for nonvegans). Examples of foods that may cause imbalance in the *kapha dosha* include berries, cherries, mango, peaches, pears, beets, cabbage, carrots, cauliflower, eggplant, radishes, spinach, sprouts, barley, corn, rice, rye, and chicken, turkey, and shrimp (for nonvegans). Both John Douillard and Gabriel Cousens include comprehensive lists of foods that are best and foods to avoid for each *dosha*.

Practitioners seek to determine the primary *dosha* and the balance of *doshas* through questions that allow them to become very familiar with the patient. The practitioner will ask about diet, behavior, lifestyle practices, and the reasons for the most recent illness and symptoms the patient had. He will carefully observe such physical characteristics as teeth, skin, eyes, and weight.

He will also take the person's pulse because each *dosha* is thought to have a particular pulse speed.

Ayurvedic Healing

Ayurvedic practitioners expect patients to be active participants in their treatment because many Ayurvedic treatments require changes in diet, lifestyle, and habits. In general, treatments use several approaches, often more than one at a time. The goals of treatment are to:

- *Eliminate impurities.* A process called *panchakarma* is intended to be cleansing; it focuses on the digestive tract and the respiratory system. For the digestive tract, cleansing may be done through enemas, fasting, or special diets. Some patients receive medicated oils through a nasal spray or inhaler. This part of treatment is believed to eliminate the cause of disease.

- *Reduce symptoms.* The practitioner may suggest various options, including yoga exercises, stretching, breathing exercises, meditation (a conscious mental process using certain techniques— such as focusing attention or maintaining a specific posture to suspend the stream of thoughts and relax the body and mind), and lying in the sun. The patient may take herbs (usually several), often with honey, with the intent to improve digestion, reduce fever, and treat diarrhea. Sometimes foods such as lentil beans or special diets are also prescribed.

- *Reduce worry and increase harmony in the patient's life.* The patient may be advised to seek nurturing and peacefulness through yoga, meditation, exercise, or other techniques.

- *Help eliminate both physical and psychological problems.* Vital points therapy and/or massage may be used to reduce pain, lessen fatigue, or improve circulation. Ayurveda proposes that there are 107 "vital points" in the body where life energy is stored, and these points may be massaged to improve health. Other types of Ayurvedic massage use medicinal oils.

Other Relevant CAM Practices

Mind-Body Medicine

Mind-body medicine focuses on the interactions among the brain, mind, body, and behavior, and on the powerful ways in which emotional, mental, social, spiritual, and behavioral factors can directly affect health. It is an approach that respects and enhances each person's capacity for self-knowledge and self-care, and it emphasizes techniques that are grounded in this approach. Mind-body medicine includes therapies like meditation, prayer, mental healing, and therapies that use creative outlets such as art, music, or dance.

Energy Medicine

Energy therapies, similar to quantum theory, involve the use of energy fields. They are of two types: veritable, which can be measured, and putative, which have yet to be measured. The veritable energies employ mechanical vibrations (such as sound) and electromagnetic forces, including visible light, magnetism, monochromatic radiation (such as laser beams), and rays from other parts of the electromagnetic spectrum. In contrast, putative energy fields (also called biofields) have defied measurement by reproducible methods. Therapies involving putative energy fields like acupuncture, are based on the concept that human beings are infused with a subtle form of energy.

Some forms of putative energy therapy manipulate biofields by applying pressure to and manipulating the body by placing the hands in or through these fields. Examples include Qi Gong, Reiki, and therapeutic touch. Vital energy is believed to flow throughout the material human body. Therapists claim that they can work with this subtle energy. Practitioners of energy medicine believe that illness results from disturbances of these subtle energies.

Light therapy is the use of light to treat various ailments. The use of light therapy has been documented to be beneficial for seasonal affective disorder, with less evidence for its usefulness in the treatment of more general forms of depression and sleep disorders. Hormonal changes have been detected after treatment. Although low-level laser therapy is claimed to be useful for relieving pain, reducing inflammation, and helping to heal wounds, strong scientific proof of these effects is still needed.

Naturopathic Doctors and Naturopathic Medicine

Naturopathic medicine is a Western-based healing practice. Many raw food leaders are or were trained as Naturopathic Doctors (NDs). Among them are Ann Wigmore and Norman Walker. There have been some observations that naturopathic medicine, provided in the four U.S.-based and two Canadian naturopathic medical schools, has been evolving toward a more conventional approach as of late. Students are exposed to a significant amount of education about nutrition and are trained to treat the cause, not the symptoms. This is consistent with the philosophy of the raw and living food lifestyle.

Currently, only 14 states license NDs. These states are Alaska, Arizona, California, Connecticut, Hawaii, Idaho, Kansas, Maine, Montana, New Hampshire, Oregon, Utah, Vermont, and Washington. NDs can practice in other states, but their practice will be limited in most cases to being a consultant, and they will not have the ability to order tests or write prescriptions. So if you live in a state that does not license NDs but choose to work with an ND for your care, you may want to make sure they have a relationship with an MD or other practitioner who can order tests and work in conjunction with the ND on your healthcare.

Six Principles of Naturopathic Medicine

Naturopathy is a school of medical philosophy and practice that focuses on improving health and treating disease chiefly by assisting the body's intrinsic ability to recuperate from illness and injury. Naturopathy practices the following six principles:

1. **First do no harm.** The process of healing includes the manifestations of symptoms, so any therapy that interferes with this natural healing process by masking symptoms is considered suppressive and should be avoided.

2. **The healing power of nature.** The healing power of nature has two aspects: first, the body has the ability to heal itself and it is the naturopathic doctor's role to facilitate this natural process; and second, nature heals. Following this principle includes getting enough sleep, exercising, proper nutrition, and, if needed, additional "earth food," such as herbs or algae (a living organism). Plants can gently move the body into health without the side effects caused by some synthetic chemicals in modern pharmaceuticals.

3. **Identify and treat the cause.** For healing to take place, practitioners must remove the underlying root causes of disease. These root causes can exist at many levels: physical, mental, emotional, and spiritual. It is the naturopathic doctor's role to identify the root cause, as well as alleviate suffering by treating symptoms.

4. **Treat the whole person.** A core tenet of naturopathy is the belief that health must go beyond treatment of immediate symptoms and instead care for the person's entire well-being. That means treating the whole body, as well as the spirit, soul, and mind.

5. **The physician as teacher.** The naturopath's role is also to educate the patient in naturopathic practices and encourage patients to "take responsibility for their own health." This cooperative relationship between doctor and patient is essential to healing.

6. **Prevention.** The ultimate goal of the naturopathic physician is prevention. The emphasis is on building health, not fighting illness. NDs do this by fostering healthy lifestyles, healthy beliefs, and healthy relationships.

Mainstream MDs with a CAM Approach

Many physicians take a CAM approach to diagnosing and treating patients, using both CAM and conventional medicine. The physicians discussed in this section are all leaders in complementary medicine. These physicians, well-known for their publications, radio shows, or other public promotion of CAM, are in great demand but have limited hours to dedicate to patient practice.

Some of the more well-known traditional physicians with a CAM approach include Dr. Ronald Hoffman, Dr. Mehmet Oz, and Dr. Andrew Weil.

Ronald L. Hoffman, MD

Dr. Hoffman has been providing natural approaches to healthcare on his *Health Talk* radio since the 1990s. In addition to his traditional medical training, he is also a licensed acupuncturist and a certified nutritionist. If all physicians were trained like Dr. Hoffman, we would likely have a better approach to healthcare overall. He summarized the most common concerns of patients in his 1997 book, *Intelligent Medicine: A Guide to Optimizing Health and Preventing Illness for the Baby-Boomer Generation.* This book was billed as "featuring the best of alternative and mainstream approaches."[8] In his 2006 book, *How to Talk with Your Doctor: A Guide for Patients and Their Physicians Who Want to Reconcile and Use the Best of Conventional and Alternative Medicine,* Hoffman cites these advantages of a CAM approach:[9]

- Patient-centered care
- Lower toxicity and fewer side effects
- Emphasizes the doctor-patient relationship
- Engenders hope
- Focuses on prevention
- Addresses concerns ignored by mainstream medicine
- Makes the patient an active participant in care (most important of all)

Mehmet C. Oz, MD

Although he has many medical accomplishments, Dr. Oz's most significant contribution to the healthcare world is his persistent determination to educate patients to take responsibility for understanding their bodies and to take better care of themselves. In addition to patient responsibility, Dr. Oz focuses on basic nutrition and self-management. While the concept of patient responsibility is not limited to CAM practitioners, he has inspired tens of millions of

Americans in a very short time. A regular guest on the *Oprah Winfrey Show*, he is likely to reach millions more.

In one of his first books, *Healing from the Heart: A Leading Surgeon Combines Eastern and Western Traditions to Create the Medicine of the Future*, Dr. Oz discusses his willingness to explore complementary approaches to traditional medicine with his patients.[10] But Dr. Oz's first big splash into the American consciousness came with the publication of *YOU: The Owners Manual: An Insider's Guide to the Body that Will Make You Healthier and Younger*, which he co-authored with Michael Roizen, MD.[11] The book is a must-read.

Andrew Weil, MD

Dr. Weil, a Harvard Medical School graduate, is probably most famous for his series of books on integrative medicine with a focus on healing. In one of his first books, *Spontaneous Healing: How to Discover and Embrace Your Body's Natural Ability to Maintain and Heal Itself*, he explores different case studies of patients told they had no hope of recovery, but somehow, with no intervention, they survived and thrived.[12] These spontaneously healed patients often had some belief system, practices, or other approaches that influenced their ability to thrive in the face of what appeared to be a fatal diagnosis. His approach is holistic and carefully thought out.

Raw and Living Food Institutes

You can find instruction in raw and living food preparation aimed at improving your health at the following locations:

> Hippocrates Health Institute, West Palm Beach, Florida
> Ann Wigmore Foundation, San Fidel, New Mexico
> Ann Wigmore Institute, Puerto Rico
> Tree of Life Rejuvenation Center, Patagonia, Arizona
> Living Food Institute, Atlanta, Georgia

CHAPTER 13

Spirituality

···

"*The immaterial essence of* an individual life or a person's total self"[1] defines spirituality. In his book on mindfulness meditation, *Wherever You Go, There You Are,* Jon Kabat-Zinn defines spirituality as "experiencing wholeness and interconnectedness directly, as seeing that individuality and the totality are interwoven, that nothing is separate or extraneous."[2] Raw and living foodists are some of the most spiritually sound people I have encountered. It may have to do with the fact that individuals who practice the principles of raw and living food have great respect for the earth and for every living thing. There is also a great appreciation for whatever great *abundance* of food they are blessed with. Generally, there is a lack of focus on the self and more focus on helping others.

As I approach my second year of practicing the living food lifestyle, I have just gotten the food thing down. In the realm of spirituality, I have recently completed meditation instruction to begin a meditation practice.

I am still in many ways on the outside looking in. I consider this helpful because it has allowed me to observe the full picture of the raw and living food lifestyle. I can only hope to someday embody some of the spiritual strength that I have witnessed in the many raw foodists I have had the pleasure to meet during the past two years.

In the past few years, opportunities to improve your spirituality have increased. For example, on a drive one day from New York to Baltimore, I listened to the Oprah and Friends radio station on XM. On that day, I heard Wayne Dyer describe his interpretation of the Lao Tzu's *Tao Te Ching*, Maryann Williamson provide insight into *A Course in Miracles,* and Jon Kabat-Zinn discuss mindfulness meditation with Mehmet Oz. While this may be spiritual overload, it speaks to a bigger issue. We all want and need to feel more meaning to the *immaterial essence of our lives*. So it appears that many of us seek to augment our current spiritual practices.

In this chapter, we will briefly explore the spiritual practices, separate from organized religion, that are common among raw foodists. We will discuss their roots in Eastern religion and philosophy. There are more similarities than differences between these practices. For example, certain yoga and meditation practices are performed as one unified whole. Many are intimately related to the Indian practice of Ayurvedic medicine, which is discussed in the chapter on health and healthcare.

Meditation

Meditation describes a state of concentrated attention, self-inquiry, and increased awareness. It usually involves turning the attention inward to a single point of reference, like breathing. Meditation is recognized as a component of most Eastern religions, where it has been practiced for over 5,000 years. During his address to the Millenium World Peace Summit at the United Nations, S. N. Goenka, the Vipassana Acharya, called meditation a nonsectarian remedy to a universal problem.[3]

Meditation is a path to inner peace. Peace and harmony, however, cannot coexist with negativity, mental corruption, or impurity.[4] You use meditation to quiet your mind and allow the negativity, corruption, and impurities to dissolve. You quiet your mind by observing your thoughts. If you observe negativity or other unsettling thoughts in your mind, they lose their strength and fade away. The more you practice observing without losing your mental balance, the easier it will be to maintain your balance when you face stress, negativity, or change.

Meditation in Buddhism

Meditation is central to Buddhism. The historical Buddha himself was said to have achieved enlightenment while meditating under a Bodhi tree. Most forms of classic Buddhist meditation distinguish between two classes of meditation practices, *shamatha* and *vipassana*, both of which are necessary for attaining enlightenment. *Shamatha* is a practice that develops the ability to focus on an outer object or an inner object. *Vipassana* is an analytic form of meditation, which does not focus on an object. Classic Buddhist meditation masters suggest that these two practices need to be developed in sequence, first *shamatha* followed by *vipassana*.

Maharishi Mahesh Yogi (commonly known as the guru to the Beatles), who introduced meditation (TM)® to the West, once said that if 1 percent of the population meditated daily, the other 99 percent would benefit from the vibration and pull of love generated by the meditators. The TM® technique makes use of quantum theory in describing the evolution and development of TM®. Today, it is estimated that .4 percent of the population meditates daily. Maharishi Mahesh Yogi created the trademarked Transcendental Meditation, or TM™, technique.

The Maharishi organization claims that since 1968, over six million people have been trained on TM®. TM® is a seven-step training process used to teach the technique that is practiced for 20 minutes twice a day. The technique is highly guarded by the Maharishi organization as well as by the students of TM®, who sign an agreement not to divulge details of their training. The Maharishi organization offers the first two steps free of charge to the public in most major cities throughout the world. After receiving the initial instruction, if you commit to the remaining five steps, which involve five two-hour sessions of training, you will pay a fee for the remaining instruction. Your fee covers the cost of the initial training program and attendance at any training sessions held by the Maharishi organization anywhere in the world, for the rest of your life.

In *Everyday Zen*, Charlotte Joko Beck describes what meditation practice is and what it isn't. Because we often have so many expectations about what we will get from meditating, Joko Beck spends time describing what meditation practice is not. For example, she says it is not about producing psychological change, achieving some blissful state, cultivating special powers or personal power, or having nice or happy feelings. She does say that meditation practice is simple, and it's about ourselves. To practice effectively, we need to remove ourselves from all external stimuli. Then we experience reality, which is challenging for most of us. Joko Beck describes this in the following way, "Our interest in reality is extremely low . . . we want to think . . . to worry through all of our preoccupations . . . to figure life out."[5] Through the meditation practice, we get acquainted with ourselves.

Mindfulness Meditation: Meditation to Heal

Jon Kabat-Zinn describes meditation as "waking up and living in harmony with oneself and the world" and mindfulness as "paying attention in a particular way: on purpose, in the present moment, and non-judgmentally."[6] In 1979, Kabat-Zinn founded the Center for Mindfulness at the University of Massachusetts Medical Center. His many books and CDs include: *Wherever You Go, There You Are; Full Catastrophe Living, Coming to Our Senses: Healing Ourselves and the World through Mindfulness; and Mindfulness Meditation*. In his Stress Reduction Clinic, Kabat-Zinn and his team have treated and healed many patients, although he will be the first to point out that this is not the purpose of the meditation practice. Healing and recovery may be side effects of meditation practice, according to Kabat-Zinn, but not the reason for engaging in the practice. "Meditation is not for the faint-hearted nor for those who routinely avoid the whispered longings of their own heart," he says.[7]

Meditation training is available in many different forms. David Lynch's book, *Catching the Big Fish: Meditation, Consciousness, and Creativity,* is an easy read and a wonderful explanation of his own personal experiences with meditation. S. N. Goenka teaches classic meditation in centers worldwide. You can obtain more information on these free 10-day highly disciplined programs at www.dhamma.org. Vipassana Fellowship teaches classic meditation online. The 90-day program costs 125 dollars. You can learn more at www.vipassana.com. Transcendental meditation is also taught in centers around the world. You can learn more about the program at www.tm.org. Finally, Mindfulness Meditation is taught in various locations each year by Jon and Saki Santorelli. You can learn more about these programs at www.umassmed.edu.

Yoga

Yoga means *union*. Although many think this term refers to union between body and mind or body, mind, and spirit, the traditional meaning is the union between one's individual consciousness and the Universal Consciousness. Therefore, yoga refers to a certain state of consciousness as well as to methods that help you reach that goal or state of union with the divine. It is a group of ancient practices of spiritual and physical development originating in India. Yoga is intimately connected to the religious beliefs and practices of Indian religions. As you will see below, one of the eight limbs of yoga practice is meditation. Outside of India, yoga is mostly associated with the practice of asanas (postures) of hatha yoga or as a form of exercise. Major branches of yoga include: hatha yoga, karma yoga, jnana yoga, bhakti yoga, and raja yoga.

The eight-limbed path forms the structural framework for yoga practice. No one element is elevated over another in a hierarchical order. Each is part of a holistic focus which eventually brings completeness to the individual as they find their connectivity to the divine. You can emphasize one branch and then move on to another as you round out your understanding in your yoga practice. The Eight Limbs of yoga practice are:

(1) *Yama* , meaning moral restraint and abstinence from attachment to possessions.

(2) *Niyama*, meaning personal discipline, purity, contentment, study, and surrender to God.

(3) *Asana*, meaning body postures used in yoga practice.

(4) *Pranayama*, meaning breath control, life force, or vital energy.

(5) *Pratyahara*, meaning control of the senses.

(6) *Dharana*, meaning concentration and cultivating inner awareness.

(7) *Dhyana*, meaning devotion and meditation on the divine.

(8) *Samadhi*, meaning union with the divine.

There are many different types of yoga. Kundalini yoga, in particular, focuses on spirituality. Gabriel Cousens dedicates significant portions of his book *Spiritual Nutrition* to the description of kundalini as the key to spiritual evolution and energy. In fact, Cousens tells us that the book is the second-to-last publication to emerge from his own meditation practice where he received kundalini initiation from Swami Mukran Paramahamsa. The practice of kundalini yoga consists of a number of bodily postures, expressive movements and utterances, meditations, breathing patterns, and degrees of concentration. Recently there has been a growing interest within the medical community to study the physiological effects of meditation, and some of these studies have applied the discipline of kundalini yoga to their clinical settings.

There are many schools local and worldwide that provide yoga classes and yoga teacher training. In the United States, the Kripalu Center for Yoga and Health in the Berkshires is probably one of the most popular. You can find out more about their programs at www.kripalu.org. Many yoga programs, like those offered at your local gym, may focus more on the physical aspect of the practice than on the spiritual. You need solid physical training to sustain a yoga practice. However, you will want to locate a program that melds the physical and spiritual components for the best experience.

Fasting

Fasting is traditionally defined as abstaining from food. When we abstain, we deliberately refrain from an activity, often with great effort. Most raw and living foodists practice fasting. The raw and living food lifestyle fasting practice may include abstaining from solid foods, all foods, media (media fasts are actually quite popular and can clear your head), or abstaining from other activities like dining out or using a computer or cell phone. The reason for fasting is twofold. Fasting cleanses the body of toxins. These toxins may be the kind that are caused by imbalanced eating. They may also be radiation emitted by computer screens or stress levels created by listening to or watching too much negative news. Fasting is probably the oldest known method for healing.

Fasting can be a spiritual practice. Jesus, Elijah, and Moses fasted for 40 days. They did not fast for physical health. They fasted for communion with God.[8] In *Spiritual Nutrition,* Gabriel Cousens describes religious 40-day fasts. He describes his own 40-day fast as a profound spiritual experience. He began the fast by drinking fresh green juices, and he ended by only drinking water. Cousens also suggests that no one should undertake such a fast without experienced supervision.

Although many religious fasts are for the symbolic 40 days, a fast does not have to be 40 days long in order to have spiritual benefits. Earlier in the chapter, spirituality was described as relating to a person's *total* self, or allowing for wholeness or interconnectedness. Spirituality pertains to the integration of body, mind, and soul. Fasting may be one of the best examples of total integration. In fasting, we improve our physical health through the cleansing process. We improve our mental health by practicing the rigorous discipline and consciousness necessary to choose day after day to continue the fasting practice. We also have the opportunity, if we choose, to improve our soul by using the quiet and concentration necessary to engage in a fast. We can do this through prayer and meditation. Many yogis believe that when we pray, we are talking to God. But when we meditate, we are listening to him.

Spiritual Nutrition by Gabriel Cousens and *Fasting and Eating for Health* by Joel Fuhrman are two excellent books that can guide you through the basics of the fasting process—long or short. In addition, both Cousens and Fuhrman offer supervised fasts in a retreat format.

CHAPTER 14

Physical Fitness

··

Just like eating your fruits and veggies, exercise should be an important part of your daily routine. Grandma's words ring true. The FDA's food pyramid even has a recommendation that we exercise at least 30 minutes per day. If your healthcare team is worth its weight in salt, they will discuss this with you. And hopefully you talk with your children, parents, and loved ones about their need to be physically fit. With all of the positive influences available to us, it seems unlikely that any of us would *not* be physically active. However, anecdotal, personal, and official evidence (like the obesity rate in the United States) point to a different reality.

There are many resources available about different ways to get and stay fit. This chapter will focus on the specific and unique ways that raw and living foodists get physically fit and stay that way. Let's look first at the recommendations of MyPyramid, the new food pyramid available at www.mypyramid.gov.

The food pyramid now contains a recommendation that everyone engage in at least 30 minutes of physical activity per day. Physical activity simply means movement of the body that uses energy. Walking, gardening, briskly pushing a baby stroller, climbing the stairs, playing soccer, or dancing the night away are all good examples of being active. For health benefits, physical activity should be *moderate* or *vigorous*. *Moderate* physical activities include walking briskly (about 3½ miles per hour), hiking, gardening and yard work, dancing, golf (walking and carrying clubs), bicycling (less than 10 miles per hour), weight training (general light workout). *Vigorous* physical activities include running or jogging (5 miles per hour), bicycling (more than 10 miles per hour), swimming (freestyle laps), aerobics, walking very fast (4½ miles per hour), heavy yard work (such as chopping wood), weight lifting (vigorous effort), and basketball (competitive). Activities that are aerobic and involve movement of the body will oxygenate the body. Activities that are anaerobic (like weight lifting) provide important benefits (like fighting osteoporosis and improving balance), but do not oxygenate the body.

Some may argue that the government does not have a place in determining what we should eat or what our exercise habits should be. While the government makes recommendations such as those on the MyPyramid site, the government, of course, does not actually *determine* our activities. And while it would be ideal for each of us to determine our own diet and exercise habits, the truth is that it is the media and corporate America that does this for us. What you see and hear on the radio, television, driving down the street, and in the supermarket are unfortunately more likely to drive our food and exercise choices than our own conscious decision making.

Of all of the reasons to engage in physical fitness, what is the most important one to you? If you have read other chapters in the book, particularly the chapters on pH balance and healthcare, you may have already included *get enough oxygen* on your list. As far back as 1931, Dr. Otto Warburg and other scientists found that the human body needs sufficient amounts of oxygen to stay healthy. Since acidic foods rob our bodies of oxygen, it helps to balance the acid-alkaline content in your body. But moving your body so your lungs can get to the point of complete inspiration is most important.

You can't oxygenate your body sitting at a desk, lounging on a couch, or driving a car. But you can get it by performing *moderate* to *vigorous* physical activity.

A second reason to exercise is to keep your lymph system in motion. Our lymph system, which is responsible for fighting infection and cleansing our entire insides, does not have its own pump. Unlike the circulatory system, which has the heart as its central pump, the lymph system has none. And what's more important is that without movement, your lymph system cannot effectively perform its cleaning and protecting functions. We were built with the assumption that we would engage in regular movement and this would keep our lymph and immune systems functioning. Certain activities, like jumping on a trampoline (believe it or not) are more efficient at moving the lymph system than others. We'll discuss the use of a rebounder (mini-trampoline) later in the chapter.

A third reason to engage in physical activity is that it helps to balance your spiritual and mental health. Yoga, described in the chapter on spirituality, is an excellent way to achieve this balance. Regular yoga practice also provides significant physical benefits. In this chapter, we will focus on the physical benefits of yoga, since we have already addressed the spiritual benefits.

Rebounding and the Immune System

Your immune system defends your body from invasion by bacteria, viruses, chemicals, cancer, and other foreign agents. The first line of defense against invasion is your skin and mucous membranes. They act as a physical barrier against unwelcome foes. The lymph system is your second line of defense. The lymph cells identify foreign cells or organisms like viruses or cancer, surround them, and destroy them, eventually dumping them into the blood to be eliminated. The lymph system is also primarily responsible for keeping the fluid moving in your body. So swelling in certain areas, especially in the lower limbs, is usually a sign of an inefficient lymph system. As noted in the beginning of the chapter, the lymph system has no pump of its own, so it relies on muscle activity and exercise for the lymph to circulate. For this reason, it is believed that lack of physical activity increases the rate of infections. The reverse is also believed to be true: exercise improves resistance to illness.[1]

Conditions or activities that *suppress* the immune system include aging, traveling (especially by plane), chemicals in the food you eat, pollutants in your environment, prescription and recreational drugs, poor eating habits, infections, lack of sleep, vitamin or mineral deficiencies, and stress. In addition to exercise, the following things are believed to *support* the immune system: amino acids, enzyme-rich diet, deep breathing, yoga, laughter, fasting, water that is living and structured (see the chapter on water), meditation, positive affirmations and attitudes, and eating fresh, organic food. You want to do whatever you can to keep your immune system happy and functioning well.

Now let's turn to the specific activity of rebounding, or jumping on a mini-trampoline. What does this have to do with the immune system? As a newcomer to rebounding and a former marathon runner, I would like to say that rebounding is a lot of *fun!* (Fun is part of keeping the immune system healthy). I never experienced the degree of lightness and happiness during a six- or seven- or eight-mile run that I do every time I jump on that little trampoline. I keep one at the entrance to my back door so if I (or anyone else in my family) need a quick pick-me-up, it's always there. If you have ever had the experience of jumping on a trampoline, perhaps you recall the carefree feeling. It is fun and hardly the kind of thing we traditionally associate with exercise that is good for you. And by rebounding, you are simultaneously engaging in two activities that support your immune system: exercise and laughter or happiness.

In her book, *Rebounding and Your Immune System,* Linda Brooks describes the essence of the rebounding process. She states that rebounding strengthens every cell of the body equally and simultaneously. The reason for this is that as you bounce, you are "accelerating and decelerating on the same plane, or vertically, *with* gravity. During other exercise such as walking or running, you accelerate and decelerate on a horizontal plane. In normal exercise you *oppose* gravity."[2] In essence, rebounding is like resistance training for each cell in your body. NASA tested and then adopted rebound training for the astronauts when it was discovered to be 68 percent more effective to the total body than treadmill running. Certified rebound instructors, like Linda Brooks, and others claim that rebounding flushes the entire lymphatic system and triples white blood cells in two minutes of bouncing a few times per day.

The actual activity of rebounding can be as soft and light or as high energy as you would like. Your immune system benefits just as much from a soft, light bounce (hardly coming off of the surface of the trampoline) as it does from a high, fast bounce. The instruction books that accompany the rebounders describe different movements that will help you increase strength, stamina, or both. But, again, you will receive significant benefit from doing a light rebound a few times a day for just *two minutes* each time.

One last comment regarding the rebounder. Many books and studies have found an interconnectedness between all of the atoms in our bodies and the atoms in the universe.[3] Quantum theory holds that there is a constant wave, or vibration, that emanates from everyone and everything. The characteristics of that vibration, or its energy, get passed along to the next atom it encounters. It seems pretty clear that energy emitted from rebounding will be positive, and this energy will positively impact all that it comes in contact with. I can attest to the fact that, although I usually felt the endorphin high after 10 or 12 miles of running, there were many times during the run itself that the energy coming from me was more negative than positive. Think about your own fitness routine—if you have one—does it make you feel positive and happy *during* the routine as well as after you are finished?

If you are interested in rebounding, there are several manufacturers who sell them, most via the Internet. I use the Needak and find it very durable. There are also deluxe models, but they may not be necessary for you. If you have balance problems, you can purchase a rebounder with a bar so you can hold it while you are bouncing.

Yoga Practice

We discussed yoga in the chapter on spirituality. Yoga, which means *union,* is an integrated practice for improving the mind, body, and inner spirit. Yoga sees the body as a vehicle for the soul in its journey toward enlightenment. There are three integrated bodies in yoga.[4] These are the physical body, the causal body (which stores subtle impressions in the form of karma), and the astral body. The mental, spiritual, physical, and emotional benefits of yoga work together. The whole (yoga) is greater than the sum of its parts.

The physical benefits of yoga. Since the body was seen as a vehicle for the soul by the ancient yogis, the system of yoga was developed specifically to ensure that the whole body would function properly. One of the functions of yoga is to keep the body in proper physical shape (as a safe haven for the soul). The physical exercises in yoga are called *asanas*. They are nonviolent and provide gentle stretching that acts to lubricate the joints, muscles, ligaments, tendons, and other parts of the body. *Asanas* help to tone the nervous system, improve circulation, release tension, and increase flexibility.[5] As a result, yoga probably provides the most complete form of exercise available.

There are different schools of yoga, each with its own focus and purpose. Hatha yoga is the overlying school of physical yoga. It stresses mastery of the body as a way of attaining a state of spiritual perfection. It is composed of gentle and fairly easy to master poses to be performed with intense concentration for concurrent meditation. Ashtanga yoga, which is a form of hatha yoga, is much more vigorous and fast paced.

The *asanas*, or positions of ashtanga yoga, can be difficult to obtain and maintain unless your body has been adequately stretched and flexed through training with basic yoga forms, like hatha. There are over 100 schools of yoga. And there are several schools of physical yoga that have grown out of hatha yoga. Some of the more common schools include bikram, kundalini, iyengar, gentle, and power yoga. Each has a specific approach to the physical postures or *asanas*.

There are 12 basic poses plus a series of poses called the *Sun Salutation* that can serve as the basis for long-term yoga practice. Yoga instruction is available not only in most towns and cities, but on DVDs that can be found in most bookstores and on the Internet. The immediate physical benefits of the yoga *asanas* (poses) include increasing the flexibility of the spine, toning and rejuvenating the nervous system, increasing flexibility of joints and muscles, and massaging the glands and organs. Circulation is improved, ensuring a rich supply of nutrients and oxygen to all cells of the body.[6] You can find out more about yoga, yoga schools, and teacher certification through the American Yoga Association at www.americanyogaassociation.org.

Walking

Walking and hiking are very beneficial activities that we often overlook when we think of the need to exercise. In fact, walking and hiking are part of the agenda for the program at Gabriel Cousens' Tree of Life Rejuvenation Center and the Ann Wigmore centers. In addition, Dr. Mehmet Oz and Dr. Michael Roizen have recently produced a CD, entitled *You on a Walk*, that is designed to walk the user through some gentle meditation as well as encouragement during a regular walking routine. The great thing about walking is that you can do it with your spouse, your children, parents, friends, and others you care about. Unlike other physical activities, you can actually focus on conversations with each other instead of keeping pace or catching your breath.

Adaptive Fitness

Like adaptive eating, you should also practice adaptive fitness, or an exercise plan that works well for you. What works well for one person may not work well for the next. This suggestion is also supported by Roger Williams' theory of Biochemical Individuality mentioned earlier in the book. Try different practices for a week or so and see how you feel. Try not to make conclusions about the benefits of an activity like yoga, rebounding, walking, or any other program you try. Your body will need at least a few weeks of engaging in the activity (three or four times a week) to determine if the exercise is working for you. Some ways to assess your progress would be to ask: Am I in pain? Do I feel overall more energetic? Do I look forward to doing the activity? Am I watching the clock while I am doing the activity? You get the idea. Your questions should be about how you feel mentally and physically. If going to yoga class stresses you out, it might not be for you. But you won't know unless you try.

It is possible to plateau and get to a point where your body gets so used to the same exercise and eating that the benefits it should provide just aren't happening as regularly. For example, if you have been practicing hatha yoga for a few years, at some point you may find that you need to practice the more challenging ashtanga postures once or twice a week. Variety makes a difference, and it can help you stretch physically and mentally.

You can apply this concept to any activity. For example, years ago, I began running because I found it was the only way for me to keep my migraine headaches at bay. First, I ran about two miles per day. After a year of this, I needed to run about three miles per day and so on. Eventually, I tapped out on the benefits of running and found other activities like rebounding and biking that took the place of running. Someday, I may run again. And, who knows what other activities I may try. There are many possibilities!

You probably don't need to consider the concept of jump-starting your exercise routine with more challenging activities until you have been involved in the same routine, providing you with benefits, for at least a few years. Even then, this does not mean that you have to abandon your routine forever, just that you may need to jump-start it. You can change your routine a bit by adding intervals—increasing speed or difficulty for a minute during each 10-minute segment that you exercise. Or take a break for awhile and try something new. Ask yourself, do I look forward to my exercise time? Do I feel happy during and after my exercise time? If your answer to any of these questions is no, it is most important for you to try new activities so you don't stop exercising altogether. It is much easier, from both an emotional and a physical perspective, to change from one routine to another than it is to start exercising again after a dormant period.

CHAPTER 15

Ethics of the Raw
and Living Food Lifestyle

· ·

The definition of ethical is, "pertaining to human character or behavior considered as good or bad with moral duty to obligation, the distinctions between right and wrong, good and evil."[1] This definition supports what we already know. The word *ethical* is subject to as many different definitions as there are people or organizations that come up with their version of what is ethical. So while many of us might say there are many good ethical outcomes of the raw and living food lifestyle, not everyone will agree.

The good, moral actions of the raw and living food lifestyle are many. First, a diet that eliminates meat saves animals, water, and energy and increases health. A diet that does not cook food saves energy and increases health. A diet that uses only organic food is kinder and gentler to the earth. But it's not just about diet. The raw and living food lifestyle embraces veganism in all areas, not just food. This means no animal products are consumed by humans, no leather shoes and no wool coats are worn, and no animal fat-based soaps are used. As Jo Stepaniak says in *Being Vegan*, from an ethical perspective, leather, dairy, and meat are indistinguishable.[2]

A second component of the ethics of the raw and living food lifestyle is the focus on peace and peacefulness. Some of this is drawn from the concentration on spirituality. We are all interconnected. Whether you look to Eastern religions, yoga, meditation, or quantum theory as the basis for peace, it begs the question, what would it be like if we could all get along? Some, like Gabriel Cousens, who believes that we create peace by being peace, practice peace on a large scale. Others just practice everyday serenity and exhibit kindness toward others they encounter. Either activity is a positive contribution.

John Robbins, whose work on food, diet, and healthcare has caused mainstream America to make some changes, is an ethics-based author and educator. What is most striking is his constant reminder to us that each of us can, every day, make choices that impact our own health and also the health of our world. It is important to understand, as people like John Robbins and Al Gore remind us, that we can make a difference, one person at a time. All ethics *is* local.

Being Vegan

The Vegan Society in England defines veganism as, "a way of living which seeks to exclude—*as far as is possible and practical*—all forms of exploitation of, and cruelty to, animals for food, clothing, or any other purpose." Vegans also use the acronym AHIMSA to describe their practice. Ahimsa is actually a Sanskrit word for nonkilling and nonharming. The six pillars of this dynamic philosophy for modern life (one for each letter: A-H-I-M-S-A) are: *A*bstinence from animal products, *H*armlessness with reverence for life, *I*ntegrity of thought, word, and deed, *M*astery over oneself, *S*ervice to humanity, nature, and creation, and *A*dvancement of understanding and truth. Being vegan is ultimately about honoring life. Being vegan is only one component of the raw and living food lifestyle. One of the primary dietary differences between veganism and the raw and living food lifestyle is that vegans eat cooked food.

In her book, *Being Vegan,* Jo Stepaniak answers many questions from readers. Her response to one question about whether it would be okay to drink the milk from a cow and eat the eggs from a chicken left abandoned on the reader's property sums up the ethics of veganism. She says, "Vegans do not view cows and birds as machines and do not believe their milk or eggs should be used as commodities. The aberrant reproductive capacities of cows and chickens have been contrived through genetic engineering, drugs, hormones, rich feed, artificial insemination, and environmental manipulation. In a natural setting, a cow would lactate only to nourish her calf until it was weaned, and chickens would produce eggs, their offspring, at a drastically reduced rate."[3]

Jo Stepaniak's book, *Being Vegan,* is comprehensive and well thought out. In the book, Stepaniak cites the Vegan Code of Ethics:

1. Choose foods that are exclusively plant-based.
2. Withhold economic and moral support from enterprises that exploit or abuse animals or humans.
3. Choose materials and products that neither destroy nor distort the lives of sensate creatures.
4. Reject the use of living creatures as instruments of materials for teaching, scientific inquiry, entertainment, or other utilitarian purposes.
5. Attempt to resolve conflict with sensitivity, respect, and nonviolent strategies.

Diet is only one component of veganism. The practice of veganism extends to all areas of life. This includes withholding support for research or any other activities that harm animals, investing only in companies that have vegan-friendly practices, and only using vegan-based personal care products, including shampoo, soap, toothpaste, laundry detergent, and fabric softener, to name a few. These products can be found in your local health food store.

Most vegans do not hold reproduction in high regard because they feel that adding more people to the planet will increase pollution and ultimately harm the environment. Besides, they say, there are many children to be cared for, and we don't need to add to the already staggering number.[4] This is one area where it appears that vegans and raw and living foodists may not be in agreement. Although I have not been able to find specific documentation of the belief in procreation on the part of raw and living foodists, many of the leaders of today's movement have children. Furthermore, many who I have spoken with are quite anxious to grow their families so that there will be a larger population born into the raw and living lifestyle, as opposed to adopting the lifestyle in adulthood.

Peacefulness

Nutritionist and author Natalia Rose puts it well when she says that when we eat animal flesh, we are ingesting their vibration, which is the fear of death. So by not eating animal flesh, we may be more likely to be peaceful people. This perspective is also supported by the vegan philosophy and ethics described above. In her book, *Raw Food Life Force Energy*, Rose describes how little in the typical grocery store is actually food, although it is packaged and sold to us as food. She cites products like processed cheese slices, heavily processed luncheon meats, mini-meals for kids, microwave fast foods, fluorescent-colored yogurt snacks that "contain so much food coloring they could double as paint," and instant soups with ingredients that "read like a radioactive experiment gone berserk."[5] As Rose points out, it is challenging to be peaceful if we are taking foreign substances into our bodies.

Gabriel Cousens provides us with a different perspective. He says that world peace is a function of our inner peace. By our state of meditation we create peace by being peaceful. Cousens believes that we all need to participate in creating peace on earth. In his book, *Living Nutrition*, he proposes that people worldwide should meditate for 40 minutes per day at sunrise or sunset, linking together thoughts of peace and visualizing the planet surrounded in light.[6]

The John Robbins' Food Revolution

John Robbins has impacted so many through his three books, *Diet for a New America*, *The Food Revolution: How Your Diet Can Help Save Your Life and Our World*, and *Reclaiming our Health: Exploding the Medical Myth and Embracing the Sources of True Healing*, he merits his own section in this chapter on lifestyle ethics. While all of Robbins' books are excellent reads, most of the information in this chapter is gleaned from his book, *The Food Revolution*.

John Robbins is a staunch proponent of the vegan lifestyle. He was heir to the Baskin-Robbins Ice Cream empire, which he abandoned shortly after his uncle, Mr. Baskin, had a fatal heart attack at the age of 52. It was at that point when Robbins started to link diet with health. Robbins began to travel a different path than the one his family followed. The thing that distinguishes Robbins from many vegans is his passion for organic and locally grown food. In chapter 5 of *The Food Revolution*, he lays out the specific ingredients for a healthy plant-based diet. Robbins believes our diets should: (1) include lots of fresh vegetables and fruits, (2) be low in refined and processed foods and sugar, (3) be free from hydrogenated fats, MSG, artificial preservatives, artificial colors, and chemical additives, (4) include more water and less soda, (5) exclude fried food, and (6) use locally and organically grown foods when possible.[7]

The facts in Robbins' book are well-researched, and it is difficult for anyone to argue successfully against the points he makes about eliminating animal-based foods from our diets. First, Robbins talks about what we feed the animals we eat. You can read the unsavory details for yourself in his book. Here is a sampling of what you will find. "Dried poultry waste and sewage sludge are routinely fed to U.S. cattle. In 1997, in the wake of the British epidemic of mad cow disease, the U.S. FDA finally banned the practice of feeding cow meat and bone meal back to cows . . . recycled chicken manure is routinely incorporated back into the diets of U.S. chickens."[8]

Another ethical issue addressed by Robbins is the direct impact on the environment created by producing various foods. He uses a table of information on how many gallons of water it takes to produce certain types of foods as a way to get us thinking about ethics and waste in our choices. For example, according to specialists at the University of California, it takes about 23 to 25 gallons of water to produce 1 pound of any of the following: lettuce, tomatoes, potatoes, and wheat. It takes 49 gallons of water to produce 1 pound of apples, 815 gallons for 1 pound of chicken, 1,630 gallons to produce 1 pound of pork, and 5,214 gallons for 1 pound of beef.

In chapter 16, Robbins addresses the biotechnology industry. He describes how Monsanto Corporation, founded in 1901 by a chemist to manufacture saccharin, is the largest player in genetic engineering. The CEO of Monsanto claims that biotechnology is the solution for a safe and sustainable tool in agriculture and nutrition, in human health, and in meeting the world's needs for food and fiber. Robbins reminds us that the top five biotechnology companies (Monsanto, Astra-Zeneca, DuPont, Novartis, and Aventis) manufacture nearly 100 percent of genetically engineered seeds.[9]

In chapter 18, Robbins cites Monsanto again. In 1999, he says, 87,000 bags of organic tortilla chips were destroyed when a routine analysis found transgenic DNA to be present in the product. The organic corn had been grown on a small organic farm, but it had been contaminated by cross-pollination from neighboring farms that were growing genetically engineered corn. Robbins describes how companies responsible for this debacle, like Monsanto, instead of apologizing, sue the farmers whose crops were contaminated, accusing them of stealing a patented product—the genetically modified version of corn.[10]

A more compelling genetically modified organism (GMO) example is what happens when human growth genes (or any growth genes) are injected into animals, or even plants. You have seen giant apples or tomatoes in the grocery store, right? Chances are great that they were genetically modified to reach that size. For example, the tomato seeds may have been genetically modified by adding fish hormones or even human hormones to produce the desired larger plant. Eating this food creates a significant set of imbalances. First, for the humans who eat the food, it creates hormone imbalances. Second, the GMO fish (or animals) mate with untainted native fish, and this gives the genetically engineered fish a selective advantage, and they create more offspring. But as it turns out, the offspring of genetically engineered fish have far higher mortality than the native fish.[11]

Robbins ends his book, appropriately, reminding us that the choices we make individually and collectively do make a difference. He says that the destiny of life on earth is up for grabs, and we are each a part of how it will turn out.

In summary, ethics, or doing the right thing, in the raw or living lifestyle turns out to be not too different from the ethics of vegans, humanitarians, environmentalists, and people who care about other people and the earth. This includes not eating animals or animal products, eating only foods that are natural, untainted, and organically grown, not using (if at all possible) any animal or animal products, treating everyone with kindness, looking for peaceful resolutions to problems, and being kind to the earth.

CHAPTER 16

The Economics of Transitioning to a Raw and Living Food Lifestyle

···

Proposing change of any kind in the 21st century requires continuous and careful thought. Politics and economic interest groups are at the heart of many of these considerations. Without the money and powerful support, sweeping change is likely not to occur. The challenge with the raw food lifestyle goes beyond politics and money. It ultimately goes to credibility and scientific support for the raw and living food lifestyle. Dr. Dean Ornish writes in his foreword to John Robbins' book, *The Food Revolution: How Your Diet Can Help Save Your Life and Our World,* "I have spent most of my professional life using the latest *high*-tech medical technology to assess the power of *low*-tech and *low*-cost interventions."[1] The types of interventions Dr. Ornish refers to include a low-fat, plant-based, whole foods diet, yoga and meditation for stress management, and exercise. We can ask, how can anyone not support a diet that is fresh, tasty, healthy, and organic? Dr. Ornish's work is a testament to the fact that the complexity of our world has made the answer to this question not so straightforward.

Today, there is a handful of raw and living lifestyle leaders who have made some significant contributions to research supporting the diet and the lifestyle. However, Dr. Ornish has one additional advantage that, to date, none of the raw food leaders can claim. Ornish was a *mainstream* physician. He was accepted as a scientist who practiced alongside his colleagues at the local hospital while he pursued his *alternative* nutrition and fitness interests. This gave Dr. Ornish a higher likelihood of gaining credibility and acceptance among his peers and society at large.

The attempt to obtain support for raw activities like eating more organic foods, dehydrating instead of cooking, relying on nuts and seeds for bulk in your diet, rebounding, and even sprouting or drinking wheatgrass, pales in comparison to what Dr. Ornish and his supporters have accomplished. Ornish conducted the studies, created the documentation, and wrote the books to communicate his findings.

Finding support for the cause is only the first hurdle. Once you have obtained credibility and have scientific proof for the raw and living food solution, you still need to be able to propose a plan for implementation that would not require the entire economic infrastructure to first fall like a house of cards and then be rebuilt. When Hillary Clinton proposed positive but sweeping changes in healthcare for the Clinton administration in 1992, the largeness of the change was so overwhelming for most Americans that it shook their sense of security. And ultimately the plan was abandoned. Interestingly, we have implemented pieces of the plan during the past 16 years under different names like HIPAA and the Medicare Patient's Bill of Rights. Most likely, in the next few years, we will look back on what was proposed by Hillary Clinton almost 20 years earlier and realize we have a program that is very similar to the one she proposed. We achieved the solution in small increments instead of all at once.

There are six important components of the raw and living food lifestyle that should be studied for scientific, political, and economic support. These include: (1) a requirement for at least *sustainable,* and ideally, *organic* food coming from all food producers in the United States, (2) *banning refined sugar* (or at least providing a warning about the negative health impact), (3) providing education on keeping *living enzymes* in at least some portion of our daily diet, (4) acknowledging the importance of eating a diet with a *healthy pH balance*, (5) improving *the quality of drinking water*, and (6) providing education on the importance of *amino acids*. In the remaining part of this chapter, we will look at the economics of implementing the six proposed raw and living food lifestyle recommendations above. This includes a potential decrease in healthcare costs, the impact on our domestic food suppliers and related industries, and the impact on current raw food suppliers (excluding fresh produce suppliers). We'll end with a look at a grassroots movement that asks, *what if* each one of us did something to make a difference every day?

The Decreased Cost of Healthcare . . . and Increased Personal Responsibility

The decreased cost of healthcare. That has a nice ring to it, doesn't it? We have been talking about this issue in the United States since Medicare and Medicaid programs were created in the mid-1960s. The costs of healthcare have escalated tremendously over the past four decades and the trend appears to be continuing. There are two characteristics about the raw and living food lifestyle that could help to reverse this trend. The first is pure, healthy eating. The second is increased individual responsibility. The issue of pure and healthy eating and its impact is pretty clear and has been addressed throughout this book. The real challenge regarding healthy eating is not acknowledging that healthy eating is good for you, but doing something about it. That is where increased responsibility, inherent in the raw and living lifestyle, will have the biggest impact regarding the economy and healthcare costs. Personal responsibility is an intrinsic part of a raw and living lifestyle. Taking increased responsibility leads to decreased healthcare costs.

The raw and living lifestyle, as described in this book, is one that requires the individual to be aware, inquiring, and engaged in critical thinking. If you are practicing the raw and living lifestyle the way it is meant to be practiced, you will be aware or conscious, and involved in a critical, analytical thought process about all of your activities—not just eating. Although the process may begin with your focus on what goes into your mouth, it will soon extend to all areas of your life. What do I use to wash my hair? Did I remember to bring my canvas bag to the grocery store so I won't have to use additional plastic or paper? Are the nuts used in this recipe sprouted, to ensure proper enzyme content? How can I be more mindful when making decisions that affect my employees? What is a kind, peaceful way to settle the disagreement I am having with my child, parent, or spouse? You get the idea. The areas for engagement and improvement are almost never ending.

Increased awareness and engagement in the raw and living lifestyle can begin with something as simple as looking for the words *conventional* and *organic* on the signs above the bananas, apples, or spinach you are about to purchase and then picking the ones with the organic sign. Eventually, you may refine this process to one where, instead of driving to the local supermarket for your fresh produce, you pick up locally grown organic produce that is delivered to a common point in your hometown. The difference between supermarket and locally grown organic produce can be demonstrated in a recent encounter I had with kiwi. It was late January and I was looking for kiwi for my son's lunch for school the next week. Thrilled that the only remaining carnivore in my family had agreed to eat fresh fruit every day, as long as it was kiwi, I had not thought about the practical reality of this request.

At the closest Whole Foods Market in eastern Pennsylvania (which for me is a 45 minute trip), I found organic kiwi and lots of them. However, when I looked closer at the tiny print on the label, I saw that it said *Italian organic* kiwi. Sure that this kiwi was not all the way from Italy and that it was instead a brand name of some sort, I asked one of the workers in the produce department who confirmed for me that yes, that kiwi had been shipped to the U.S. from Italy. Now, my dilemma was twofold. First, I imagined the trip that kiwi had taken to make its way from the warmth of southern Italy, across the Atlantic, through the docks in New York City, and then across the highway to eastern Pennsylvania. All I could think was, "How could this piece of fruit have *any* energy, let alone nutritional content, left in it?" Second, what exactly does *organic* mean in Italy? I did not have my laptop computer with me, so I couldn't Google the term *criteria for organic produce in Italy*, but I imagined that Europeans are very kind to their food, so organic in Italy was probably at least as sound as organic in the U.S. I bought the kiwi with the plan that I would stop at the three other supermarkets I would pass on the way home to see whether they had *organic* kiwi that was *domestic*. The supermarkets were Fresh Market, Giant Foods, and Super Fresh. I guessed that it was highly unlikely that I would find a better quality of kiwi at any of these stores. I was correct. Only two of the stores even had kiwi. And, of those two stores, both had conventional kiwi and both were imported. Given the choices available, I felt that I had made the right decision. While the result was not optimal, at least my son would be eating imported *organic Italian* kiwi and not Oreos or hot wings.

Some degree of mindfulness is necessary when you practice a raw and living lifestyle—a degree that is not necessary if you are on the standard American diet (SAD). A decade ago, if my son had asked for kiwi, I would have added it to the list and mindlessly plopped them into the cart (without knowing what country it came from or whether it was organic or conventional) with 20 other items purchased from the cookie and cereal aisle while I was thinking about the meeting I had the night before. I usually had my cell phone to my ear during these shopping trips and I was more focused on the conversation than on the food I was selecting. Those days are over for me, and they can be for you too if you are willing to take on the responsibility. The rewards are great.

The second part of increased responsibility is the relationship between an individual who assumes increased personal responsibility and the corresponding improvement in her healthcare. There are two striking examples of this and both are supported by empiric, published studies. In his book, *Reclaiming Our Health*, John Robbins describes the studies performed at Yale University by Bernie Siegel, the noted surgeon and author. Dr. Siegel and his colleagues found a 100 percent correlation between the head nurse's opinion of the patient and the long-term survival rates. If the head nurse said, "He's a real s.o.b. and won't let you draw the blood for the test," Dr. Siegel's team would find no trouble with the patient's immune system. If the patient was a submissive, gentle patient who would not question anything, and would always let the nurses draw blood, he was in trouble.[2] One of the possible interpretations of the results of these studies is that patients who take responsibility for themselves and are willing to *stick their neck out* are more likely to get better results. You can also interpret this action as representative of someone who is aware, engaged, and critically thinking about the impact of someone else's actions on him.

In 1997, the results of a psychological study of nursing home residents were published.[3] 91 residents were split into three groups. The first group received a speech from the director of the home emphasizing that the residents had a lot of responsibility for their own lives. She also told them that a movie would be shown on two nights during the upcoming week and that the residents had to decide on which night they would attend. The director gave each resident a house plant, stating that it was up to the resident to take care of it. The director also gave a speech to the other (comparison) group, but she omitted all references of taking responsibility and making decisions for themselves. This group was assigned to a movie night, and each person was also given a plant. However, this group was told that a nurse would care for the plant. The final group was assigned a movie night and had no speech or plant given to them, just business as usual in the nursing home for them.

When the researchers conducted a follow-up study 18 months later, they found that the intervention not only improved residents' health but reduced the likelihood that they would die. After 18 months, 15 percent of the experimental group had died compared with 30 percent in the comparison group. The findings suggest that giving elderly people more responsibility and control over their lives and their environment may slow down mental and physical decline. More importantly, there are possible implications for every one, regardless of age. Not being actively involved in the decision making about your life and your day-to-day activities can have a detrimental effect on your health and well-being. The converse, being more aware, mindful, and engaged in decision making about your life, can have a positive effect on your health and well-being.

Proposed Impact of Raw Lifestyle Changes on U.S. Farmers and Manufacturers

The impact on the economy of the six proposed changes at the beginning of this chapter for the economy could be staggering. Let's look at each one individually.

First, phasing out conventionally grown crops and replacing them with sustainable and eventually all organic crops. John Robbins, in *The Food Revolution*, discusses the fact that many studies have found that the amount of food produced from organic fields are comparable to conventional (non-organic) processes. Robbins includes a nod to the Rodale Institute. In 1995, Rodale found that during the first 14 years of their 15-year study, comparable yields were obtained without the use of chemical pesticides or fertilizers. Even more significant was the fact that they found that organic fields produced more product than conventional fields during drought years.[4]

Second, eliminating refined sugar from the American diet or at least including a warning on the label. The economic impact of such a concept is far beyond the scope of this book. If the tobacco industry is at all analogous, the amount of time it will take to: (1) prove the harm caused by refined sugar, (2) identify ways for the industry to repair the damage done, and (3) identify suitable substitutes industry-wide could be quite lengthy. The current sugar industry could take control by beginning to manufacture and distribute healthy, *natural* sugar substitutes like agave, stevia, or yacon. Unless there is a significant economic incentive, the sugar industry is highly unlikely to undertake such actions.

Third, providing information on the importance of eating foods with their enzymes still intact. It appears that sufficient work has been done to prove the value of living enzymes in food. How and whether industry, the government, and perhaps more importantly, the media address this issue is an open-ended question. In order for enzymes to get the attention they deserve, there would have to be something in it for each of the players. Industry could benefit through production and sales of either supplements or tools for use in the sprouting process. Government could benefit by spending less on healthcare because of the resulting healthier population. And, ultimately, if the American public benefits, at least in theory, everyone should benefit.

Fourth, making the pH of all packaged foods part of the information on the package ingredient label. This should be a given since pH is such an important component of every food. It is probably more important than the number of calories on the current label. The initial cost to calculate the pH level may be high, but once it is calculated, the requirement should be easy to regulate.

Fifth, improving the quality of drinking water. This is probably a state or local issue overall. But sales of bottled water could be regulated at the federal level. And sixth, increasing amino acid rich foods. This could be addressed somewhere in the U.S. Department of Agriculture's food pyramid. There is much more that needs to be done to move toward a set of healthier habits for all of America. Yes, we can, and we have, moved forward individually. But, when we have the backing of all relevant groups, we will all be healthier, and then we can focus on the many other pressing issues that deserve our attention besides dieting and eating.

The Supply and Demand of Raw and Living Food

This section addresses the supply and demand of prepared raw and living food. Ideally, the majority of food preparation should be done at home. However, at least in the short term, it will likely be impossible to change the American mind-set regarding the ability to grab something on the way to or from work or some other planned activity. Even more important, what about the ability to purchase convenience foods so you have a complete meal, not just salad and fruit? There are already some raw food merchants providing this service. They include Matt Amsden of RAWvolution (www.rawvolution.com). He also offers prepared, vacuum-packed, raw food that can be delivered to your door each week. You receive that week's four entrées, two desserts, four side dishes, and two soups. This is enough to feed one person dinner for about four days. Certain foods preserve better than others and you might not like everything on the predetermined menu, but overall, the experience is a pleasant one.

Another raw merchant who delivers food overnight is Awesome Foods. From the Web site at www.awesomefoods.com you can order any number of entrees, delicious raw breads, soups, salads, and desserts. As long as you order by Sunday at 5 p.m., you will receive your food the following week. This organization specializes in raw *carbohydrate-based* foods and is particularly helpful for the transitioning raw foodist or for the winter months when you may need food that is more substantive. Their raw breads come in flavors like onion, olive, and rye, and instead of the crunchiness so common in raw breads (which makes them seem more like crackers), they are a thin, soft, chewable texture— great for making sandwiches with Awesome Foods' *better than chicken salad.*

In addition to mail order fast raw food, most larger cities and some smaller towns have raw food cafés that offer quick sit-down meals or takeaways. My family and I regularly use Pure Food and Wine takeaway and Quintessence takeout, both in New York City. Each restaurant has its own unique style and selections. Both are excellent. Whenever I travel, I do a quick Google search for *raw food cafés* in the city I will be visiting. Today, there is about a 75 percent chance that there will be a café in the selected city I am headed for. A year and a half ago, the percentage was more like 50. This may be an indication of increased demand in a short time.

These examples provide some information about the current supply and demand for raw food. Other ways to determine growth in demand is to look at the growth in the number of Web sites dedicated to selling raw foods. A year ago, when I did a Google search on *raw chocolate bars*, one or two sites would show up on the return page that were valid suppliers of raw chocolate candy bars *and* were U.S.-based companies. At that time, most companies supplying raw chocolate were in Europe. Today, there are at least six suppliers in the U.S. Even then, there is a good chance that for one or two of those companies, the chocolate bars will be out of stock, which further indicates that demand is outpacing supply.

Raw ice cream is another example of a supply and demand imbalance. As I noted earlier in the book, high-quality raw ice cream (made from organic young Thai coconut meat, cashews, and agave) is one of the tastiest and most fun raw foods. However, the challenge to making it is twofold. First, the supply of high-quality young Thai coconuts is tenuous at best. Second, if you can find the coconut supply, you still need able-bodied people to harvest the coconut pulp from the body of the coconuts—a daunting task. Sarma of Pure Food and Wine makes the most amazing raw ice cream on the market. Hers is creamy and consistent in quality and taste. Sarma is a perfectionist and will not make a product unless it is sure to please the customer *and* meet her standards. In the fall of 2007, when supply of resources and coconuts were not adequate, Sarma stopped serving ice cream in her Pure Food and Wine takeaway. You could almost hear the groans in New York City's Gramercy Park section when she made this decision. But Sarma took the time to create a workable strategy to meet demand. By late winter 2008 (a grueling four months later), Sarma had the process humming again with ice cream available at her takeaway store and from her online store (www.oneluckyduck.com).

These are all anecdotal pieces of information about supply and demand. But they indicate a trend and a concern. If we want to encourage a raw and living diet, one of the best ways to get and *keep* people on it is by providing choice and excellent taste in food. The many leaders in the raw food preparation movement noted in this book have proven that taste and choice are possible. What is still yet to be proven is how it is possible to make high-quality raw and living foods so the consumer of the food retains the benefits of organic, freshness, high enzyme content, *and* love. The attitude of the individual making the food carries vibrations that impact whoever consumes that food. Raw food preparation staff are aware of this and take their jobs and their responsibilities seriously. But unless we can train thousands of people a year in this process (keep in mind that they need to be the right people), there is an issue of demand outstripping the supply of raw food. Somehow, the idea of mass-producing raw chocolate bars—all of which, by the way, are now made by hand—removes an essential ingredient.

There is no quick fix. As Einstein said, "In the middle of difficulty lies opportunity." With the growing demand for raw foods, there appears to be an opportunity. Now we just have to meet it in a way that does not turn the purity and healthiness of raw foods into some perverted version of what it used to be. We need to keep raw and living foods just that—raw and living.

A Grass Roots Movement for Raw and Living Food: Things We Can Do

10 Things We Can All Do

Start out by selecting two from the list below to help make yourself healthier and the world a better place.

1. Eat some greens everyday.
2. Eat less meat, or none at all.
3. Eat no refined sugar or products containing refined sugar.
4. Jump on a rebounder 20 minutes a day or practice yoga daily.
5. Fast once or twice a year from food or media.
6. Look up the pH level of the 10 most common foods you eat and eliminate anything with a pH level less than 6.
7. Drink only bottled water from artesian wells (like Fiji).
8. Eliminate dairy or only use raw dairy products.
9. Meditate 20 minutes a week.
10. Pick one super supplement to add to your daily regimen: bee pollen, blue-green algae, chlorella, or spirulina as one of your daily supplements.

10 Things Schools Can Do

1. Provide organic lunches.
2. Serve lunches made with produce grown locally.
3. Educate students about the importance of pH balance.
4. Educate students about the food industry.
5. Train teachers on the relationship between nutrition and health.
6. Raise awareness about the relationship between nutrition and health.
7. Invite speakers like John Robbins and other socially conscious pioneers from private industry to give presentations.
8. Use healthy sweeteners like stevia and agave instead of sugar and artificial sweeteners in cafeteria food.
9. Teach an elective course on meditation (Harlem public schools teach meditation to their students).
10. Teach yoga during gym class—for boys and girls.

10 Things Private Industry Can Do

1. Require all employees to participate in a company-sponsored nutritional training program.
2. Feed employees fresh, natural, organic, and/or raw food in the cafeteria.
3. Serve blueberries and almonds at company meetings instead of coffee and donuts.
4. Provide incentives to employees to eat well (contests or recognition programs).
5. Require all employees to take company-sponsored mind-body stress reduction training.
6. Stock the company cafeteria with healthy sweeteners like stevia and agave instead of sugar and artificial sweeteners.
7. Offer regular yoga classes first thing in the morning or during lunch breaks.
8. Keep rebounders in common areas.
9. Invite John Robbins and others to speak.
10. Provide quality herbal tea alternatives with agave or stevia at the coffee machine.

10 Things the Government Can Do

1. Add pH content to food labels.
2. Require teaching nutrition (and its role in health) to grade school and high school students (state governments).
3. Pass legislation to certify Naturopathic Doctors to practice in the 36 states that do not currently allow NDs to practice.
4. Post health warnings on refined sugar and refined sugar products.
5. Pass green legislation—more oxygen means less acidity.
6. Highly regulate and eventually phase out the nonorganic farming industry.
7. Require schools to serve organic options for lunch.
8. Provide more resources and training on the food pyramid (www.mypyramid.com).

9. Update the food pyramid annually with input from all segments of the population, not just industry. This would include input from private nutritionists, physicians, and alternative medicine practitioners.
10. Pilot all of these suggestions at one (or more) schools in each school district. After a year, measure improvements in health, grades, and physical fitness.

CHAPTER 17

Pets and Raw Food

· ·

If you Google the search term Raw Diet, you will find that about 2 of every 10 entries are for raw *pet* diets. For example, one of the first 10 entries is for the barfworld.com Web site. The entry states, "Information about the Bones and Raw Food Diet, with recommendations and sale of foods, supplements, publications and videos." When you click on the entry, you land on a Web site dedicated to the raw diet of pets that is quite impressive and very comprehensive. The tabs include entries for BARF association, BARF breeders, and BARF distributors. The acronym BARF means *biologically appropriate raw food* or *bones and raw food*. On the BARF diet tab, the first paragraph states, "Your pet deserves only the most nutritious, safe, healthy and natural raw pet food on the planet. BARF World's philosophy is that in order to achieve optimum health for your pet from puppy to senior you should feed a *biologically appropriate raw food* or *BARF* diet which is rich in vitamins, minerals, living enzymes, natural protein sources and premium quality fruits and vegetables."[1] See also, barfaustralia.com for more information.

It's not just the Internet that is full of resources on raw diets for pets. There are also many books written by very experienced, credible individuals who provide advice and guidance on the highly recommended raw diet for pets. Some of the most noted books include *The BARF Diet* by Australian veterinarian Ian Billinghurst who uses the BARF diet for his clients' pets and for his own pets. Denise Flaim is the companion-animal columnist for *Newsday*. Flaim's book is entitled, *The Holistic Dog Book: Canine Care for the 21st Century.* Other books include *Raw Dog Food* by Carina Beth MacDonald, *Natural Nutrition for Dogs and Cats* by Kymythy R. Schultze, and *Dr. Pitcairn's Complete Guide to Natural Health for Dogs and Cats* by Richard H. and Susan Pitcairn.

With the exception of animal flesh (which is actually a pretty big exception), the raw food pet diet is not that dissimilar from the human raw and living diet described in this book. There is a middle-ground solution for those of us without the stomach or the conscience to serve our pets raw animal flesh. This middle ground is proposed by Denise Flaim, and we will address it in the section on Raw Animal Ethics below. In this chapter, we will discuss the history and the components of the raw pet diet, as well as the ethics of it.

The Roots of the Raw Pet Diet

Dr. Ian Billinghurst is the pioneer of the BARF diet. He was the first to publish a complete resource on the diet describing his own pets' experiences with it as well as the experiences of the animals he has treated. The acronym BARF stands for *biologically appropriate raw food* or *bones and raw food,* depending on which one appeals to you more. Dr. Billinghurst claims that the diet brings enormous health benefits to animals that eat it. Most other supporters of the diet who have written their own books or have Web site resources on the diet cite Billinghurst as the source for many of their own practices.

Billinghurst explains that BARF is about "feeding dogs and cats the diet they evolved to eat over millions of years of natural, species-appropriate genetic adaptation . . . simple logic and basic biology dictates that because our pets' bodies require an evolutionary diet, that is exactly what we should feed them."[2] He explains in more detail that the further away an animal's diet is from its natural, evolutionary diet, the more health problems that animal is likely to develop. As a result, he is convinced that modern grain-based dog feed causes many health problems. BARF, Billinghurst explains, is merely a return to the biologically appropriate method of feeding pets that was abandoned about 65 years ago when *processed* pet foods came about. Processing of food is not only not good for us, it's not good for our pets either.

The genetic difference between domesticated dogs and gray wolves is about 1 percent. This minor degree of difference between dogs and their wolf kin is the key component of the nutritional theory supporting raw dog food. Dogs, like wolves, need raw food to derive crucial enzymes and nutrients, which are destroyed during the cooking process.[3] Sound familiar? This last sentence, with few modifications, could be included in the introduction to this book in its explanation about why raw food is best suited for humans.

Benefits of the BARF Diet

Dr. Billinghurst says that, as a general rule, any genetic faults your pet may have inherited will have a minimal chance of expression when you feed them the BARF diet. He includes the following benefits as those he personally has witnessed when pets have been transitioned to the raw diet. The pet has: (1) increased energy; (2) increased lean body mass; (3) many dental and skin problems disappear; (4) infected ears become healthy again; (5) arthritis, incontinence, diabetes, and reproductive system problems disappear; and (6) resistance to parasites increases. Flaim adds the following benefits of the raw diet to her list: (1) enhanced immune system and reduced allergic reactions; (2) increased hydration; (3) less of a doggy smell; and (4) smaller volume of stools.

Components of the Raw Pet Diet

According to Billinghurst and other BARF experts, there is a certain way to feed your pet to get maximum benefit from the raw diet. He lists the following five ingredients as the basis of the BARF diet:

1. *Water.* Pets do not get enough water when we feed them dehydrated pet food. Billinghurst suggests supplementing the animals' diet with water from a healthy source, like spring water.
2. *Raw, Meaty Bones.* Bones were a major food source in the evolutionary diet of our animals. They supply the bulk of the pets' energy, water, protein, vitamin, mineral, and enzyme requirements.
3. *Raw Fruits and Vegetables.* Dogs need approximately 15 percent of their diet as fresh, whole, raw mostly nonstarchy vegetables and fruit. Cats need about 5 percent of the diet as such.
4. *OFFAL.* Offal, or internal animal organs are important to the diet as well.
5. *Supplements.* Supplements commonly include healthy oils, probiotics, minerals, vitamins, kelp, alfalfa, garlic, and phytonutrients. Billinghurst's section on supplements is rather detailed, so if you have an interest in feeding your pet BARF, you will want to purchase his book.

Ethics of the Raw Pet Diet

In *Being Vegan*, Jo Stepaniak addresses the issue of the *official* vegan perspective on feeding meat to animals. First, she says that there is no true *official* vegan opinion and that the community is conflicted. She says that dogs are able to eat a wide variety of foods and, with proper supplementation, may be able to do well on a predominately plant-based diet. She also notes that cats require meat to thrive. She does note that vegans often cite the anatomical differences between herbivores and carnivores as the reason why humans should be vegan. Cats are in fact designed to eat animal flesh.[4] Stepaniak concludes by saying that from a moral perspective we must look at what is in the best interest of the animals in our custody, even if the solutions are less than ideal given our own belief system.

Denise Flaim, in *The Holistic Dog Book,* describes a middle ground that may be workable. She talks about the anxiety that some pet owners experience when they are confronted with zealots who insist that it's their way or the highway. It seems that the concept of biochemical individuality applies to pets as well as their owners.[5] At least that appears to be the path recommended by Monica Segal, author of *K9 Kitchen, Your Dogs' Diet: The Truth Behind the Hype.* Segal found that dogs are highly individualistic. While one dog may thrive on raw food, others may need cooked food. Some dogs may take well to certain meats, while others do not.[6] Trying different combinations until you arrive at a solution that works for your dog's health and your conscience is your best strategy.

Conclusion

···

A great deal comes under the umbrella of the raw and living food lifestyle.

Depending on your current habits and your perspective, the concept can be overwhelming. There are two ways to approach the information presented to you in this book. First, you can apply the information that appeals most to you. If you are not sure where to begin, I offer some suggestions below. Second, you can *plunge right in* and apply as many of the suggestions as possible. This approach can be overwhelming for even the most daring, so I have provided recommendations for this tactic as well.

If you would like to apply the concepts in moderation, I have listed below 10 actions you may want to take. You can incorporate one or two concepts initially, and as these become part of your daily routine, you can add a few more. If you are already practicing two or more of the concepts, you may wish to apply the other seven or eight simultaneously. For those of you who are not currently practicing any of the ideas, you might want to start with two or three of the suggestions. Over time, another step you can take is to increase or broaden the activity. For example, you can choose number four, decreasing your refined sugar intake by 50 percent. In a month or two, you may want to decrease your refined sugar intake by 75 percent and eventually your refined sugar intake may be zero. You can create your own 12-month plan by using the information from each of the chapters.

Activities you can undertake if you are approaching the raw and living food lifestyle in moderation include:

1. Creating a nutritional plan.
2. Eating more fresh, organic fruits and vegetables (some of them raw) each day.
3. Drinking fresh vegetable juice (preferably made from greens) at least once a week.
4. Decreasing current refined sugar intake by 50 percent.
5. Looking up the pH of at least one food you eat per day. If the pH is less than 5, don't eat it.
6. When purchasing bottled water, select water from pure sources like artesian wells or springs over those that are labeled *drinking water,* since the latter is probably just bottled tap water.
7. Exercising vigorously three times per week for 30 minutes each day.
8. Doing a media or a shopping fast one day per month. This includes online media and online shopping!
9. If you have never taken a yoga or a meditation class, try taking one. Then incorporate one of them into your activities once a month, or once a week if you prefer.
10. Researching complimentary and alternative healthcare providers in your area and either attending a presentation they are giving or visiting them. Ask questions and analyze why you either would or would not consider using the practitioner.

Activities you can undertake if you are approaching the raw and living food lifestyle in a more intense way include:

1. Creating a complete nutritional VVMS and referring back to it at least monthly to see if you are on the path you created for yourself. If not, determine whether you need to modify your VVMS or your actions and do so.

2. Eating at least 60 percent of your diet as fresh, organic, uncooked fruits and vegetables.

3. Eating uncooked or dehydrated seeds, nuts, sprouts.

4. Drinking fresh pressed vegetable juices daily.

5. Practicing the vegan lifestyle as much as is practicable.

6. Engaging in a food fast at least once per month.

7. Eating 80 percent of your foods with a neutral or alkaline pH level.

8. Meditating daily. If you do not currently have a meditation practice, try using one of the resources in the fitness chapter to learn more and develop a practice.

9. Drinking only water from pure and structured sources. If you must drink tap water, using crystals to restructure the water for a few hours prior to drinking.

10. Practicing a daily fitness routine that includes yoga at least four times per week.

As time passes and the new practices in your life become established practices, remember that change is good. Take a brief break every once in a while. Try something different or change the quality or quantity of an activity for a short time. The use of fasting provides a good example. If you keep doing the same thing over and over again, your body will adapt—no matter how good it is for your body—because you are an adaptive organism. What was once a challenge and interest to your metabolic process will become boring and easy. So every once in awhile, you need to do something to jump-start your metabolism. Fast more frequently, or for a different period of time, or do a water fast. There are other approaches you can take as well. For example, you can modify your fitness routine or increase or decrease the percentage of fruits and vegetables you eat daily by 10 percent. You can even take a brief vacation from one or more activities. Remember, you are your own best experiment. By constantly being aware of your daily habits and determining which principles in this book work or don't work for you—and making appropriate changes—you will likely lead a healthier and happier life. Good luck and have fun!

Glossary

. . . .

Acidic foods—foods that contain little to no oxygen, including all sweets, grains, meats, fish, eggs, milk products, animal fats, coffee, tea, unripe fruits, and alcohol.

Agave—is a sweetener commercially produced in Mexico and the southwest United States, from several species of agave, a cactus-like plant.

Ahimsa—nonviolence.

Alkaline foods—fruits, vegetables, greens, and sea vegetables.

Amino acids—the building blocks of proteins and intermediaries in the body's metabolism.

Ayurvedic medicine—integrates the body, mind, and spirit to prevent and treat disease and to promote wellness, and is one of the world's oldest medical systems. In Ayurvedic philosophy, people, their health, and the universe are all thought to be related. Health problems can result when these relationships are out of balance. The goal of Ayurvedic medicine is to cleanse the body of substances that can cause disease and to re-establish harmony and balance.

Bee pollen—is the most complete food found in nature, according to some. It helps to support the immune system, is high in vitamins A, B, C, and E, minerals, proteins, amino acids, hormones, and enzymes. Bee pollen has also been shown to have a protective effect against radiation. Gram for gram, pollen contains an estimated five to seven times more protein than meat, eggs, or cheese. In addition, the protein in pollen is in a predigested form and is easy to assimilate into the digestive system.

Biochemical individuality—Roger Wiliams' theory; based upon the fact that everyone has unique and specific, genetically determined nutrition requirements.

Bioflavonoids—these are a class of water-soluble plant pigments. They contain anti-inflammatory, antihistaminic, and anti-viral agents. They block the "sorbitol pathway" that is linked to many symptoms of diabetes. A type of phytonutrient, bioflavonoids also protect blood vessels and reduce platelet aggregation (acting as natural blood thinners).

Blue-Green Algae (also known as E3 Live algae or crystal manna)—A supplement that is harvested from Klamath Lake in Oregon. It is high in protein, chlorophyll, and vitamins, and it enhances both the immune system and brain function as it is naturally high in neurotransmitters.

Cacao—Raw seeds from the Theobroma cacao tree. They can be processed and cooked to create cocoa from which chocolate is made.

Camu-Camu (Myrciaria dubia)—is a bush native to the South American rainforest. It is higher in minerals than any other Amazonian plant. A fresh berry, it contains 30 to 60 times more vitamin C than an orange. The powder is typically found in vitamin C supplement capsules. The Camu-Camu berry is a source of phosphorus, calcium, potassium, iron, and the amino acids serine, valine, and leucine. Small amounts of the vitamins thiamine, riboflavin, and niacin are also present.

Carbon filter—The second of three steps in ideal water purification. Removes organic and inorganic chemicals that the ROI unit cannot.

Carob—Pods that grow on an evergreen shrub and are processed into chocolate-like powder for use as a sweetener in recipes.

Carotenoids—A type of phytonutrient that is a natural fat-soluble pigment found principally in plants, algae, and photosynthetic bacteria, where they play a critical role in the photosynthetic process.

Chakra—A spinning sphere of bioenergetic activity from the head to the base of the spine.

Chi—Chinese word that refers to *life force* or *life energy.*

Chlorella—high-protein algae with about 5 grams of protein in one teaspoon. It also contains high amounts of magnesium and the super detoxifier chlorophyll. It contains two to five times the amount of chlorophyll as spirulina. Chlorella is the best algae for pulling heavy metals out of the system.

Cocoa—Cocoa is the dried and partially fermented fatty seed of the cacao tree from which chocolate is made.

Crystal Energy—a detoxifier and a patented form of minerals, rich in silica. The manufacturer of the substance claims that Crystal Energy helps remove heavy metals from the body. They also state that it challenges the symptoms of dehydration and minimizes the process of aging.

D3—Vitamin D is best obtained directly from sunshine. However, in certain climates during the winter months, this is not possible. Vitamin D supplements may be necessary during those times. A natural, organic (nonsynthetic) version of vitamin D is the best.

Dehydration—the process of heating foods at a maximum temperature of 100–120° F. This preserves digestive enzymes and increases the life of the food.

Dosha—In Ayurveda, three qualities called *doshas* form important characteristics of the individual's constitution and body type. Practitioners of Ayurveda call the *doshas* by their original Sanskrit names: *vata, pitta,* and *kapha.* Each *dosha* is made up of one or two of the five basic elements: space, air, fire, water, and earth. Each *dosha* has a particular relationship to body functions and can be upset for different reasons. A person has her own balance of the three *doshas,* although one *dosha* usually is prominent. *Doshas* are constantly being formed and reformed by food, activity, and bodily processes. Everyone has a dominant dosha, and Ayurvedic practitioners use the individual's dosha to treat the patient and design nutritional programs.

Energy therapies—similar to quantum theory, involve the use of energy fields. They are of two types: veritable, which can be measured, and putative, which have yet to be measured. The veritable energies employ mechanical vibrations (such as sound) and electromagnetic forces, including visible light, magnetism, monochromatic radiation (such as laser beams), and rays from other parts of the electromagnetic spectrum. In contrast, putative energy fields (also called biofields) have defied measurement by reproducible methods. Therapies involving putative energy fields are based on the concept that human beings are infused with a subtle form of energy. This vital energy or life force is known under different names in different cultures.

Enzyme—biomolecules in the alimentary tract that break down food so that the organism can absorb it.

Enzyme inhibitors—present in seeds and nuts; can interfere with the body's enzyme production. Sprouting the seeds/nuts is said to reduce enzyme inhibitors.

Essenes—the first religious sect (second century BC) to describe, embrace and practice fasting and eating raw foods for health.

Fasting—abstaining from food for spiritual, physical, or mental reasons for a period of time. Fasting heals the body and rids the body of toxins.

Flavonoids—are polyphenoids with potential beneficial effects on human health. They have been reported to have antiviral, antiallergenic, antiplatelet, anti-inflammatory, antitumor, and antioxidant activities.

Food Combining—a principle which claims that the combination of different foods eaten at one meal can either help or hinder digestion and overall health.

Fruitarian—the practice of a diet that excludes everything but fruit and certain vegetables.

Glycemic load—gives a relative indication of how much a serving of food is likely to increase your blood sugar level.

Glycosides—sugar molecules that help remove toxic waste from the body.

Juicing—process of separating the liquid from the fiber, consuming only the liquid from the fruit or vegetable.

Kirlian photography—photos taken by Seymour and Valentina Kirlian by a machine that they adapted which measures energy patterns.

Kundalini yoga—a system of meditative techniques and movements within the yogic tradition that focuses on psycho-spiritual growth and the body's potential for maturation. The practice of kundalini yoga consists of a number of bodily postures, expressive movements, and utterances, meditations, breathing patterns, and degrees of concentration.

Living foods—foods which still have live enzymes circulating in them.

Maca—a vaguely sweet, off-white powder. Maca root grows in the mountains of Peru. It is a radishlike root vegetable that is related to the potato family. A raw sweetener.

Meditation—describes a state of concentrated attention, self-inquiry, and increased awareness. It usually involves turning the attention inward to a single point of reference, like breathing. Meditation practice helps develop your relationships with yourself and others. It is recognized as a component of most eastern religions, where it has been practiced for over 5,000 years.

Mesquite—a tan-orange, sweet and pungent-smelling powder. Mesquite bushes grow pods. Mesquite powder is then ground from the mesquite pods. High in protein, calcium, magnesium, potassium, iron, and zinc, with a sweet, molasseslike taste.

MSM—is a naturally occurring supplement with its roots in oceanic phytoplankton. Methylsulfonylmethane (MSM) is a sulfur-based naturally occurring compound and the closest chemical relative to MSM is dimethyl sulfoxide (DMSO). Available as a capsule or crystal to be taken orally. In addition, it is available in cream or lotion.

Nano B Complex—a liquid that contains living, not synthetic, B vitamins to support the liver, immune system, heart, brain, and mood balance. The substance contains vitamin B_1, B_2, B_3, B_5, B_6, B_{12}, folic acid, para-amino acid, and biotin. This product is made by Premier Research Labs. Individuals on a vegetarian or vegan diet cannot obtain their B vitamins from food sources since they occur in meats or animal-based foods. Although the body has some ability to manufacture vitamin B_{12} and other B vitamins naturally, not everyone's body can assimilate these B vitamins well, so the Nano B complex can help give vegans the advantage they need in this area. Max Stress B Nano-Plex is a type of Nano B complex.

Naturopathic Medicine—Western-based healing practice which focuses on the cause of the ailment, not the symptoms.

NCD (also called Zeolite)—NCD stands for natural cellular defense. It is an energy substance made from a compound that pulls out pesticides, radiation, metals, and other chemicals from the body. It has been shown to have a protective effect against diseases like hepatitis C. Zeolites are extracted from natural volcanic rocks. They have a crystalline shape like honeycomb that helps trap and remove the heavy toxins and metals from the cells in your body in a safe, natural way. It is only one of the few minerals that are negatively charged. As a result, zeolites apparently attract and draw the harmful metals and toxins in your body to it, capture them, and help your body remove them.

Oligofructose—a naturally occurring sugar, which the human body does not metabolize.

Ozonated—water that has had unstable oxygen molecules added to it. This destroys the frequency pattern and the water is no longer living, as it loses most of its structure in this process.

Panchakarma (Ayurveda)—A cleansing process; it focuses on the digestive tract and the respiratory system. For the digestive tract, cleansing may be done through enemas, fasting, or special diets. Some patients receive medicated oils through a nasal spray or inhaler. This part of treatment is believed to eliminate worms or other agents thought to cause disease.

pH—the *potential of hydrogen* or the number of hydrogen ions in a substance. The higher the number, the more acidic the substance. The lower the number, the more alkaline the substance. A pH of 7 is neutral, neither acidic nor alkaline.

Phytonutrients—are the category of components of plants thought to promote human health. Subcategories include carotenoids and flavonoids. Fruits, vegetables, grains, legumes, nuts, and teas are rich sources of phytonutrients.

Phytoplankton—the component of plankton that produces organic substances from inorganic molecules and the sun—also known as photosynthesis. Through photosynthesis, phytoplankton is responsible for 50 percent of the oxygen present in the earth's atmosphere. As a supplement, phytoplankton is bottled in a way that preserves the marine phytoplankton in suspended animation, protecting the original life energy. The high chlorophyll content of phytoplankton also increases oxygen uptake.

Polyphenols—a group of chemical substances found in plants. Research indicates that polyphenols may have antioxidant characteristics with potential health benefits similar to those in berries, olives, and other fruits and vegetables.

Predigestion—process by which enzymes in raw foods digest their own ingredients, thereby reducing the work of the human body's digestive system.

Probiotics—dietary supplements containing beneficial bacteria. They are live microorganisms that when administered in adequate amounts confer a health benefit on the host. Claims are made that probiotics strengthen the immune system to combat allergies, excessive alcohol intake, stress, exposure to toxic substances, and other diseases.

Pure Synergy—a powder compound made from organic vegetables, fruits, and herbs. It contains 62 different ingredients, including spirulina, Klamath Lake algae, chlorella, salina, kelp, Irish moss, dulse, wheatgrass, kamut, and Chinese rejuvenation herbs. It also contains several medicinal mushrooms.

Quartz Crystal—The final step of the three-step process in ideal water purification. The hexagonal structure of the quartz crystals will rearrange the water to its original structure. Crystals have their own particular vibration of a precise and measurable intensity. This vibration attunes itself to human vibration better than any other gem or mineral. It can also attune itself to the vibration of substances like water. Quartz crystals are used to amplify, clarify, and store energy.

Raw foods—food that is grown and eaten in its uncooked form.

Reverse Osmosis (ROI)—The first of three steps in ideal water purification. ROI removes bacteria, viruses, nitrates, fluorides, sodium, chlorine, particulate matter, heavy metals, asbestos, organic chemicals, and dissolved minerals (part of the ideal three-step process for water purification).

Spirulina—a complete protein and is about 60 to 65 percent amino acids. Contains B vitamins, minerals, chlorophyll, and phytonutrients. One of the highest sources of gamma-linolenic acid (GLA)—an essential fatty acid (EFA) in the omega-6 family that is found primarily in plant-based oils).

Sprouting—involves soaking seeds, nuts, or beans for several hours until the food begins to develop a *tail-like* protrusion on one end. The sprouting of seeds neutralizes or inactivates the enzyme inhibitors. The *good* enzyme activity is at its height when the *tail* or the *sprout* is approximately ¼ of an inch long.

Stevia—an herb used as a sweetener in raw recipes; ideal for those with diabetes because of its low glycemic load. It can be used in many forms, including powder, liquid, fresh leaf, or dried leaf.

Sustainable Agriculture—an integrated system of plant and animal production practices that have a site-specific application and that will, over the long-term, (1) satisfy human food and fiber needs, (2) enhance environmental quality and the natural-resource base upon which the agricultural economy depends, (3) make the most efficient use of nonrenewable resources and on-farm resources and integrate, where appropriate, natural biological cycles and controls, (4) sustain the economic viability of farm operations, and (5) enhance the quality of life for farmers and society as a whole.

Veganism—six principles include: (1) abstinence from animal products, (2) harmlessness with reverence for life, (3) integrity of thought, word, and deed, (4) mastery over oneself, (5) service to humanity, nature, and creation, and (6) advancement of understanding and truth.

Vegetarianism—is the practice of a diet that excludes all animal flesh, including poultry, game, fish, shellfish, or crustacea, and slaughter by-products.

Vitalzym X—helps boost enzyme production and is good for inflammation and injury. It boosts the immune system and has an anticancer effect. It is a blend of enzymes and nutrients formulated for optimal energy.

Yacon—a sweet, tuberous root vegetable that produces rich, sweet, brown syrup.

Yoga—means *union*. Although many think this term refers to union between body and mind or body, mind, and spirit, the traditional meaning is the union between one's individual consciousness and the Universal Consciousness. Therefore, yoga refers to a certain state of consciousness as well as to methods that help one reach that goal or state of union with the divine. It is a group of ancient practices used for spiritual and physical development that originated in India.

Raw and Living Food
Un-cookbooks

··

Angel Foods: Healthy Recipes for Heavenly Bodies—Cherie Soria

*Eating in the Raw: A Beginner's Guide to Getting Slimmer, Feeling Healthier, and
Looking Younger the Raw-Food Way*—Carol Alt

Eating Without Heating: Favorite Recipes from Teens Who Love Raw Food—Sergei
Boutenko and Valya Boutenko

I Am Grateful: Recipes and Lifestyle of Café Gratitude—Terces Engelhart and Orchid

LifeFood Recipe Book: Living on Life Force—Annie Jubb and Dr. David Jubb

Living Cuisine: The Art and Spirit of Raw Foods—Renee Loux Underkoffler

*Living Foods for Optimum Health: A Highly Effective Program to Remove Toxins and
Restore Your Body to Vibrant Health*—Brian R. Clement and Theresa Foy
Digeronimo

Living in the Raw: Recipes for a Healthy Lifestyle—Rose Lee Calabro

Living on Live Food—Alissa Cohen

Naked Chocolate—David Wolfe and Shazzie

Raw Family: A True Story of Awakening—Victoria, Igor, Sergei, and Valya Boutenko

*Raw Food Life Force Energy: Enter a Totally New Stratosphere of Weight Loss, Beauty,
and Health*—Natalia Rose

Raw Food Made Easy for 1 or 2 People—Jennifer Cornbleet

Raw Food/Real World: 100 Recipes to Get the Glow—Matthew Kenney and Sarma Melngailis

Raw Food Gourmet—Gabrielle Chavez

Raw Foods for Busy People: Simple and Machine-Free Recipes for Every Day—Jordan Maerin

Raw: The Uncook Book: New Vegetarian Food for Life—Juliano Brotman and Erika Lenkert

RAWvolution: Gourmet Living Cuisine—Matt Amsden

The Complete Book of Raw Food: Healthy, Delicious Vegetarian Cuisine Made with Living Foods—Lori Baird and Julie Rodwell

The Living Foods Lifestyle—Brenda Cobb

The Raw Food Detox Diet: The Five-Step Plan for Vibrant Health and Maximum Weight Loss—Natalia Rose

The Raw Truth: The Art of Preparing Living Foods—Jeremy Safron

UnCooking with Jameth and Kim: All Original, Vegan & Raw Recipes & Unique Information About Raw Vegan Foods—Jameth Sheridan and Kim Sheridan

Web Sites

· ·

Site address	What for?
Awesomefoods.com	Meals delivered, order by Sunday at 5 p.m.
Bluecanoe.com	clothes
Cafegratitude.com	Raw food café, raw food Web store, clothing and supplies
Carolalt.com	Resources, recipes
Chimachine4u.com	Chi machines and portable infrared hothouse sauna
Discountjuicers.com	Teflex sheets
Dynamicgreens.com	Wheatgrass frozen in pods—delivered in overnight mail from family-owned organic farm in Canada
e3live.com	Blue-green algae supplements, face creams, kelp soap
genefitnutrition.com	Young untreated Thai coconuts
Gnosischocolate.com	12 different types of raw, organic chocolate bars.
Goldminenaturalfoods.com	Nama shoyu, ume vinegar, tamari, tea pots, ume plums, tea tree dental floss

Goodhumans.com	Clothes, supplies
Goraw.com	Bars, chocolate, seeds, granola—Living Light Culinary Institute
Healthforce.com	Supplements
HighVibe.com	Raw foods, raw chocolates, snacks, largest selection of books, beauty products, appliances, food grown supplements, free receipes, detox programs, and much more.
Hippocratesinst.org	Raw food retreats, treatment for acute and chronic conditions, fasting, training
Kingarthurflour.com	Madagascar vanilla bean paste, The Baker's Catalogue
Kookiekarma.com	Cookies, granola, and cereal
Live-live.com	Snacks, supplements
Livesuperfoods.com	Chocolate, supplements, snacks
Livingintentions.com	Go Nuts! Seasoned nuts, other nut products including chocolate covered nuts
Livingnutz.com	Nuts
Lotusbrands.com	Teas, spices, herbs
Lovestreetlivingfoods.com	Chocolate, nuts, raw food preparation supplies
Matterofflax.com	Flaxseed crackers and hummus
Mindfully.org	Green living resources
Mindfulnesstapes.com	Jon Kabat-Zinn's mindfulness meditation CDs
MotherNature.com	Vitamin D3
Mountainroseherbs.com	Fresh, organic herbs and spices
Mrsmeyers.com	Laundry and cleaning supplies, soap
Naturalmedicine.com	Teas, herbs, list of medical disorders with natural treatment suggestions
NaturalZing.com	Snacks
NATuraw.com	Snacks
Norwalkjuicers.com	Green juicers (industrial strength)
Oneluckyduck.com	Snacks
Organicindiausa.com	Tulsi teas, Ayurvedic supplies
Palmbeachnutrition.com	Probiotics
phisciences.com	Crystal energy

Rawbakery.com	Cookies, brownies, cakes (recommended by Natalia Rose)
Rawfamily.com	Victoria Boutenko's site
Rawfoodchef.com	Classes, consultation, training, videos, Living Light Culinary Institute
Rawfoodexplained.com	Web site with information on raw food and nutrition theory
Rawforlife.com	Information, recipes
Rawganique.com	Clothes, snacks, staples
Rawguru.com	Raw ice cream! Chocolate covered bananas, Living vitamin B Nano-Plex, zeolite, infrared saunas (walk-in), ceramic knives, nut milk bags, cleavers to open coconuts
Rawvolution.com	Meals delivered weekly if ordered by Saturday
Ruthshempfoods.com	Hemp seeds
Southrivermiso.com	Chick pea and Azuki bean miso
Sproutrawfood.org	Snacks—lemon squares are the best
Sunfood.com	David Wolfe products, green superfoods, cacao, etc.
Sunorganicfarm.com	Nuts, seeds, oils, spices
Superfoodsnacks.com	Raw cacao and goji berry treats
Therawdiet.com	Sauerkraut crock
Thewolfeclinic.com	Vitamin O, Vitalzym, Far infrared ray products
Tomberlies.com	Raw vegan ice cream shipped overnight from California
Treeoflife.nu	Gabriel Cousens' training programs, snacks, some supplies, baking needs, retreats
Ulimana.com	Raw chocolate products
umassmed.edu	Center for Mindfulness homepage at UMASS Medical School
Wholespice.com	Asafoetida (Hing)

MyPyramid Nutrition and Fitness Principles

··

U.S. Food Pyramid . . . Leavin' It All Up to You

Even the government recognizes the individuality of nutritional needs. They changed the structured food pyramid to one that is now called *MyPyramid.* MyPyramid.gov is the Web site that contains nutritional resources for each of us to build our own personalized food pyramid. The site allows you to design your own plan using *Steps to a Healthier You* criteria and guidance. Just as with any resource, you need to analyze the credibility and appropriateness of the information for you. One thing you need to be aware of is the influence behind this site. The Web site contains valuable information and useful tools. However, don't take everything on face value. The government, by necessity, is influenced by every industry that affects the food market. Use the information on the MyPyramid.gov site as general guidance. Your primary criteria should be what is nutritious and what works best for you.

Let's talk about what the MyPyramid can do for you. First, it provides a basic structure that illustrates the importance of some degree of variety in your diet. While it makes suggestions about proportions of types of foods (based on the size of the color in the triangle) it does not give absolute percentages. You need to determine this for yourself. Second, it stresses the importance of certain foods. In particular, leafy greens, orange colored vegetables, and whole grains are specifically suggested in the key. Third, the pyramid gives us a "vegetarian" option by changing the category that was previously dedicated to meat and poultry to a new designation, "Meat & Beans." Fourth, the pyramid gives us a dairy-free option. If we cannot tolerate dairy products, we are encouraged to use lactose-free products instead. And, fifth, healthy cooking options are stressed. These include the suggestion to bake, broil, or grill our meats.

The figure below is the new food pyramid introduced by the federal government in 2005. This pyramid may create more questions than answers. What it undoubtedly does is leave the decision making up to you. The site provides tools for you to input your age, gender, and activity level and will make different suggestions about what you should eat. But even then, the suggestions are fairly general and will require you to make the final decision about what you eat and how much.

MyPyramid.gov
STEPS TO A HEALTHIER YOU

MyPyramid Key

Color	Food group	Sample comments from the MyPyramid.gov site
Orange	Grains	eat *whole* grains
Green	Veggies	dark green and orange
Red	Fruits	variety, including fresh, limit juices
Yellow	Oils	most from fish, nuts, vegetable oils
Blue	Milk	low or nonfat . . . if you cannot tolerate dairy products, go lactose-free
Purple	Meat & Beans	low fat/lean, vary choices . . . bake, broil, grill

The following topics are specifically addressed on the MyPyramid.gov Web site. These provide additional insight into the philosophy and the rationale behind the new pyramid. It is important to remember the origin of the *MyPyramid* is the U.S. Department of Agriculture. The origin shapes the philosophy. And as with any government-designed program, there are many interest groups looking to be kept happy. So the designers needed to consider cattle farmers, dairy farmers, bean growers, and many others in finalizing this plan. The influence of interest groups should not mean that we automatically dismiss the credibility or usefulness of the tool. Trust your analytical skills and don't take everything at face value. More importantly, use the information you know about yourself and what works for you, and apply that to the food pyramid to make it relevant to you.

Topics Addressed by the USDA on the MyPyramid.gov Web Site

One Size Doesn't Fit All

USDA's new MyPyramid symbolizes a personalized approach to healthy eating and physical activity. The symbol has been designed to be simple. It has been developed to remind consumers to make healthy food choices and to be active every day. The different parts of the symbol are described below.

Activity

Activity is represented by the steps and the person climbing them, as a reminder of the importance of daily physical activity.

Moderation

Moderation is represented by the narrowing of each food group from bottom to top. The wider base stands for foods with little or no solid fats or added sugars. These should be selected more often. The narrower top area stands for foods containing more added sugars and solid fats. The more active you are, the more of these foods you can fit into your diet.

Personalization

Personalization is shown by the person on the steps, the slogan, and the URL. Find the kinds and amounts of food to eat each day at the MyPyramid.gov. site. The theory of biochemical individuality is similar to the focus on individuality of MyPyramid.

Proportionality

Proportionality is shown by the different widths of the food group bands. The widths suggest how much food a person should choose from each group. The widths are just a general guide, not exact proportions. Check the Web site for how much is right for you.

Variety

Variety is symbolized by the six color bands representing the five food groups and oils of the MyPyramid. This illustrates that foods from all groups are needed daily for good health.

Gradual Improvement

The slogan: *Gradual Improvement* suggests that individuals can benefit from taking small steps to improve their diet and lifestyle each day.

Notes

Introduction

1. *Merriam-Webster's Collegiate Dictionary, 11th edition* (Springfield: Merriam-Webster, Inc., 2007).
2. S.E. Frost Jr., *Basic Teachings of the Great Philosophers* (New York: Anchor Books, 1989).
3. Edmond Bordeaux Szekely, *Essene Gospel of Peace: Book One* (British Columbia: International Biogenic Society, 1981), 36.
4. AskDrSears, Sears Family Pediatrics, http://www.askdrsears.com/html/4/T044900.asp.
5. Natalia Rose, *Raw Food Life Force Energy* (New York: HarperCollins, 2007).
6. Natalia Rose, *The Raw Food Detox Diet* (New York: HarperCollins, 2005).
7. Colin T. Campbell and Thomas M. Campbell, II, *The China Study: The Most Comprehensive Study of Nutrition Ever Conducted and the Startling Implications for Diet, Weight Loss and Long-term Health* (Dallas: Benbella Books, 2005).
8. Colin T. Campbell, "America's Honest Nutrition Researcher," *Living Nutrition* (2005): volume 17.
9. Gabriel Cousens, *Spiritual Nutrition: Six Foundations for Spiritual Life and the Awakening of Kundalini* (Berkeley: North Atlantic Books, 2005), 340.
10. Gabriel Cousens, *Conscious Eating* (Berkeley: North Atlantic Books, 2000), 234.
11. Gabriel Cousens, *Spiritual Nutrition: Six Foundations for Spiritual Life and the Awakening of Kundalini* (Berkeley: North Atlantic Books, 2005), 197.

Chapter 1

1. Renee Loux Underkoffler, *Living Cuisine: The Art and Spirit of Raw Foods* (New York: Avery Publishing, 2003).

Chapter 2

1. *Oxford English Dictionary, Sixth Edition* (New York: Oxford University Press, 2007).
2. Roger J. Williams, *Biochemical Individuality: The Key to Understanding What Shapes Your Health* (New Canaan: Keats Publishing, 1998), 190.
3. Preamble to the Constitution of the World Health Organization as adopted by the International Health Conference, New York, 19 June-22 July 1946; signed on 22

July 1946 by the representatives of 61 States (Official Records of the World Health
Organization, no. 2, 100) and entered into force on 7 April 1948. The definition
has not been amended since 1948.

4. Jack LaLanne, BeFit Enterprises, http://www.jacklalanne.com.

5. Lance Armstrong and Sally Jenkins, *It's Not about the Bike: My Journey Back to Life*
(New York: Berkley Publishing Group, 2001), 1.

Chapter 3

1. Susannah and Leslie Kenton, *The New Raw Energy* (London: Random House,
1994).

2. Max Gerson, *A Cancer Therapy: Results of Fifty Cases* (San Diego: The Gerson
Institute, 1997), 9.

3. Herbert M. Shelton, *Getting Well* (Whitefish: Kessinger Publishing, Inc, 1917), 14.

Chapter 4

1. *Oxford English Dictionary, Sixth Edition* (New York: Oxford University Press,
2007).

2. Edward Howell, *Enzyme Nutrition* (New York: Penguin Putnam, 1985), 141.

3. Gabriel Cousens, *Conscious Eating* (Berkeley: North Atlantic Books, 2000), 528.

4. Edward Howell, *Enzyme Nutrition* (New York: Penguin Putnam, 1985), 120.

5. Whole foods are defined as foods and food ingredients that (1) are as close to
their natural state as possible; (2) have not been highly processed or refined; (3)
have not had anything taken away or added to them to decrease their quality; (4)
are nutritionally dense and health enhancing; (5) include whole grains, legumes,
vegetables, fruits, nuts, and seeds. See Victoria Laine, *Health by Chocolate*
(Edmonton: Owl Medicine Books, 2008).

6. *Oxford English Dictionary, Sixth Edition* (New York: Oxford University Press,
2007).

7. Sally Fallon, *Nourishing Traditions: The Cookbook that Challenges Politically Correct
Nutrition and the Diet Dictocrats* (Washington, DC: New Trends Publishing, Inc.,
1999).

8. Jo Stepaniak, *Being Vegan: Living with Conscience, Conviction, and Compassion*
(Lincolnwood: Lowell House, 2000).

9. Masaru Emoto, *The Hidden Messages in Water* (New York: Atria Books, 2001).

10. Natalia Rose, *Raw Food Life Force Energy* (New York: HarperCollins, 2007).

11. Natalia Rose, *Raw Food Life Force Energy* (New York: HarperCollins, 2007),
31.

12. Harmony Health, Harmony Health Natural Therapies Clinic,
http://www.harmonyhealth.net.au/quantum.php.

13. Fritjof Capra, *The Tao of Physics* (Boston: Shambhala Publications, Inc., 1991),
68.

14. William Arntz, Betsy Chasse, and Mark Vicente, *What the Bleep Do We Know!?*
(Deerfield Beach: HCI, 2005), 56.

15. N. W. Walker, *Become Younger* (St. George: Norwalk Press, 1949).

16. Gabriel Cousens, *Conscious Eating* (Berkeley: North Atlantic Books, 2000), 298.

17. Gabriel Cousens, *Conscious Eating* (Berkeley: North Atlantic Books, 2000).
18. Natalia Rose, *Raw Food Life Force Energy* (New York: HarperCollins, 2007), 59.
19. Gabriel Cousens, *Conscious Eating* (Berkeley: North Atlantic Books, 2000), 550.
20. Natalia Rose, *Raw Food Life Force Energy* (New York: HarperCollins, 2007).
21. Hippocrates Health Institute, Hippocrates Health Institute, http://www.hippocratesinst.org.

Chapter 5

1. David and Shazzie Wolfe, *Naked Chocolate* (San Diego: Maul Brothers Publishing, 2005), 29.
2. David Wolfe's Sunfood Nutrition, Sunfood Nutrition, http://www.sunfood.com/b2c/ecom/ecomEnduser/items/xt_itemDetailNF.aspx?siteId=1&itemNum=0878.
3. Bitter chocolate is a high magnesium food. See Mildred Seelig, *The Magnesium Factor* (New York: Avery Publishing, 2003), 265. In addition, David Wolfe and Alissa Cohen of naturesfirstlaw.com, integrativenutrition.com, nakedchocolate.com, nutria-info.co.uk, and detoxyourworld.com all claim that raw cacao beans have a significant amount of magnesium. The author has been unable to find claims to refute this. The author has also been unable to find scholarly published journals attesting to this claim either.
4. Mildred Seelig, *The Magnesium Factor* (New York: Avery Publishing, 2003).
5. Health thru Natural Science, Nutritional Data, Inc. and NT Business Solutions, http://www.nutritionaldata.com.
6. Richard Beliveau, *Foods to Fight Cancer* (New York: DK Publishing, 2007).
7. Victoria Laine, *Health by Chocolate* (Edmonton: Owl Medicine Books, 2008).
8. Carl Keen, Roberta R. Holt, Patricia I. Oteiza, Cesar G. Fraga, and Harold H. Schmitz, "Cocoa Antioxidants and Cardiovascular Health," *Keen Lab—University of California, Davis*, (2005), http://keenlab.ucdavis.edu/articles/Keen529ajcn05.pdf.
9. Catherine Anne Rauch, "Chocolate: A Heart Healthy Confection?," *CNN*, (2000), http://archives.cnn.com/2000/HEALTH/diet.fitness/02/02/chocolate.wmd/#0.
10. Winifred Rosen and Andrew T. Weil, *From Chocolate to Morphine: Everything You Need to Know about Mind-Altering Drugs* (New York: Houghton-Mifflin, 2004), 10.
11. Anita A. Johnston, *Eating in the Light of the Moon: How Women Can Transform Their Relationship with Food Through Myths, Metaphors, and Storytelling* (Carlsbad: Gurze Books, 1996), 47.
12. Winifred Rosen and Andrew Weil, *From Chocolate to Morphine: Everything You Need to Know about Mind-Altering Drugs* (New York: Houghton-Mifflin, 2004), 15.
13. Jennifer Cornbleet, *Raw Food Made Easy* (Summertown: Book Publishing Company, 2005), 39.
14. The Glycemic Load is the most practical way to apply the Glycemic Index to dieting and is easily calculated by multiplying a food's Glycemic Index (as a percentage) by the number of net carbohydrates in a given serving. Glycemic Load gives a relative indication of how much that serving of food is likely to increase your blood sugar levels. Health thru Natural Science, Nutritional Data, Inc. and NT Business Solutions, http://www.nutritionaldata.com.
15. Health thru Natural Science, Nutritional Data, Inc. and NT Business Solutions, http://www.nutritionaldata.com.

Chapter 6

1. William Dufty, *Sugar Blues* (New York: Warner Books, 1975).
2. Health thru Natural Science, Nutritional Data, Inc. and NT Business Solutions, http://www.nutritionaldata.com.
3. Gabriel Cousens, *Conscious Eating* (Berkeley: North Atlantic Books, 2000).
4. Gabriel Cousens, *Conscious Eating* (Berkeley: North Atlantic Books, 2000), 611.
5. Gabriel Cousens, *Rainbow Green Live-Food Cuisine* (Berkeley: North Atlantic Books, 2005).
6. Vegetus.org, Vegan and Vegetarian Humor and Information, http://www.vegetus.org.
7. George Mateljan, *The World's Healthiest Foods* (GMF Publishing, 2006).
8. Natural News.com, Natural Health, Natural Living, Natural News; Natural News Network, http://www.newstarget.com/021506.html.
9. PERUherbals, Peruherbals, Inc., http://www.peruherbals.com/3030/yacon.html.

Chapter 7

1. Masaru Emoto, *The Hidden Messages in Water* (New York: Atria Books, 2004), 23.
2. Gabriel Cousens, *Conscious Eating* (Berkeley: North Atlantic Books, 2000).
3. Carlton Hazelwood, *A View of the Significance and Understanding of the Physical Properties of Cell-Associated Water* (New York: Academic Press, 1979).
4. Susan Brown and Larry Trivieri Jr., *The Acid Alkaline Food Guide* (Garden City Park: Square One Publishers, 2006).
5. Gabriel Cousens, *Conscious Eating* (Berkeley: North Atlantic Books, 2000).
6. Llewellyn Encyclopedia, Quartz Crystals, http://www.llewellynencyclopedia.com/article/272.
7. Barbara Hendel and Peter Ferreira, *Water and Salt: The Essence of Life* (Gaithersburg: Signature Book Printing, 2003).
8. Barbara Hendel and Peter Ferreira, *Water and Salt: The Essence of Life* (Gaithersburg: Signature Book Printing, 2003).

Chapter 8

1. Barbara Hendel and Peter Ferreira, *Water and Salt: The Essence of Life* (Gaithersburg: Signature Book Printing, 2003).
2. Gabriel Cousens, *Conscious Eating* (Berkeley: North Atlantic Books, 2000).
3. Gabriel Cousens, *Conscious Eating* (Berkeley: North Atlantic Books, 2000).
4. Susan Brown and Larry Trivieri Jr., *The Acid Alkaline Food Guide* (Garden City Park: Square One Publishers, 2006), 19.
5. Natalia Rose, *Raw Food Life Force Energy* (New York: HarperCollins, 2007).
6. Dr. Otto Warburg, "The Prime Cause and Prevention of Cancer" (lecture delivered to Nobel Laureates at Lindau, Lake Constance, Germany, June 30, 1966).
7. The Living & Raw Food Diet, Alissa Cohen, http://www.alissacohen.com.

Chapter 9

1. Fact Monster: Online Almanac, Dictionary, Encyclopedia, and Homework Help; Information Please, http://www.factmonster.com/ipka/A0775714.html.
2. Gabriel Cousens, *Conscious Eating* (Berkeley: North Atlantic Books, 2000).

Chapter 10

1. E3Live, the All-Organic Feel Good Food; E3live/Vision, http://www.E3live.com.
2. Gabriel Cousens, *Conscious Eating* (Berkeley: North Atlantic Books, 2000).
3. David Wolfe, "Superfoods" (handout given to accompany speech delivered to Institute for Integrative Nutrition, New York, New York, November 24, 2007).
4. Gabriel Cousens, *Conscious Eating* (Berkeley: North Atlantic Books, 2000), 612.
5. Michael F. Holick, *The UV Advantage: The Medical Breakthrough that Shows How to Harness the Power of the Sun for Your Health* (Ibooks, Inc., 2007), 157.
6. "Health and Nutritional Properties of Probiotics in Food Including Powder Milk with Live Lactic Acid Bacteria," report of a *Joint FAO/WHO Expert Consultation on Evaluation of Health and Nutritional Properties of Probiotics in Food Including Powder Milk with Live Lactic Acid Bacteria*, FAO/WHO, 2001.

Chapter 12

1. The Ann Wigmore Foundation: Sharing the Living Food Lifestyle, The Ann Wigmore Foundation, http://www.wigmore.org.
2. Dr. Otto Warburg, "The Prime Cause and Prevention of Cancer" (lecture delivered to Nobel Laureates at Lindau, Lake Constance, Germany, June 30, 1966).
3. Charlotte Gerson and Beata Bishop, *Healing the Gerson Way: Defeating Cancer and Other Chronic Diseases* (Carmel: Totality Books, 2007), 152.
4. Norman W. Walker, *The Natural Way to Vibrant Health* (Prescott: Norwalk Press, 1972).
5. Colin T. Campbell and Thomas M. Campbell, II, *The China Study: The Most Comprehensive Study of Nutrition Ever Conducted and the Startling Implications for Diet, Weight Loss and Long-term Health* (Dallas: Benbella Books, 2005).
6. NCCAM, National Center for Complementary and Alternative Medicine, http://www.nccam.nih.gov/health/ayurveda/#dosha.
7. John Douillard, *The 3-Season Diet: Eat the Way Nature Intended* (New York: Three Rivers Press, 2000).
8. Ronald L. Hoffman, *Intelligent Medicine: A Guide to Optimizing Health and Preventing Illness for the Baby-Boomer Generation* (Wichita: Fireside, 1997).
9. Ronald L. Hoffman, *How to Talk with Your Doctor: The Guide for Patients and Their Physicians Who Want to Reconcile and Use the Best of Conventional and Alternative Medicine* (Basic Health Publications, 2006).
10. Mehmet Oz, *Healing from the Heart: A Leading Surgeon Combines Eastern and Western Traditions to Create the Medicine of the Future* (Plume, 1999).
11. Michael F. Roizen and Mehmet Oz, *YOU: the Owner's Manual: an Insider's Guide to the Body that Will Make You Healthier and Younger* (New York: HarperCollins, 2005).
12. Andrew Weil, *Spontaneous Healing: How to Discover and Embrace Your Body's Natural Ability to Maintain and Heal Itself* (New York: Ballantine Books, 1995).

Chapter 13

1. *Random House Dictionary* (New York: Random House, 1997).
2. Jon Kabat-Zinn, *Wherever You Go, There You Are* (New York: Hyperion, 1994), 264-265.

3. Bill Higgins, "S.N. Goenka Addresses Peace Summit," *New York Times*, August 29, 2000.
4. Vipassana Meditation, S.N. Goenka, http://www.dhamma.org/en/art.shtml.
5. Charlotte J. Beck, *Everyday Zen: Love and Work* (San Francisco: HarperCollins, 1989), 26.
6. Jon Kabat-Zinn, *Wherever You Go, There You Are* (New York: Hyperion, 1994), 3-4.
7. Jon Kabat-Zinn, *Coming to Our Senses: Healing Ourselves and the World through Mindfulness* (New York: Hyperion, 2005), 22.
8. Gabriel Cousens, *Spiritual Nutrition* (Berkeley: North Atlantic Publishers, 2003).

Chapter 14

1. Elson M. Haas and Levin Buck, *Staying Healthy with Nutrition: The Complete Guide to Diet and Nutritional Medicine* (Berkeley: Celestial Arts Publishing, 2006).
2. Linda Brooks, *Rebounding and Your Immune System: Optimal Support for Your Body's Natural Defense* (Eaton: Vitally Yours Press, 2003), 29.
3. You can learn more about these studies and research. See Marlee Matlin, Elaine Hendrix, John Ross Bowie, and Robert Bailey Jr, *What the Bleep Do We Know!?* (Lord of the Wind Films, LLC, 2004). Michael J. Losier, *Law of Attraction: the Science of Attracting More of What You Want and Less of What You Don't* (Hachette Book Group, 2003). Richard P. Feynman, *Six Easy Pieces* (New York: Perseus Publishing, 2005).
4. Sivananda Yoga Vedanta Centre, *Yoga Mind and Body* (New York: DK Publishing, 1998).
5. Sivananda Yoga Vedanta Centre, *Yoga Mind and Body* (New York: DK Publishing, 1998).
6. Sivananda Yoga Vedanta Centre, *Yoga Mind and Body* (New York: DK Publishing, 1998).

Chapter 15

1. *Oxford English Dictionary, Sixth Edition* (New York: Oxford University Press, 2007).
2. Jo Stepaniak, *Being Vegan: Living with Conscience, Conviction, and Compassion* (Lincolnwood: Lowell House Publishing, 2000).
3. Jo Stepaniak, *Being Vegan: Living with Conscience, Conviction, and Compassion* (Lincolnwood: Lowell House Publishing, 2000), 27.
4. Jo Stepaniak, *Being Vegan: Living with Conscience, Conviction, and Compassion* (Lincolnwood: Lowell House Publishing, 2000).
5. Natalia Rose, *Raw Food Life Force Energy* (New York: HarperCollins, 2007).
6. Gabriel Cousens, *Spiritual Nutrition: Six Foundations for Spiritual Life and the Awakening of the Kundalini* (Berkeley: North Atlantic Books, 2005).
7. John Robbins, *The Food Revolution: How Your Diet Can Help Save Your Life and Our World* (San Francisco: Conari Press, 2001).
8. John Robbins, *The Food Revolution: How Your Diet Can Help Save Your Life and Our World* (San Francisco: Conari Press, 2001), 208.
9. John Robbins, *The Food Revolution: How Your Diet Can Help Save Your Life and Our World* (San Francisco: Conari Press, 2001), 309.

10. John Robbins, *The Food Revolution: How Your Diet Can Help Save Your Life and Our World* (San Francisco: Conari Press, 2001), 347.

11. John Robbins, *The Food Revolution: How Your Diet Can Help Save Your Life and Our World* (San Francisco: Conari Press, 2001), 350.

Chapter 16

1. John Robbins, *The Food Revolution: How Your Diet Can Help Save Your Life and Our World* (San Francisco: Conari Press, 2001), xiii.

2. John Robbins, *Reclaiming Our Health: Exploding the Medical Myth and Embracing the Source of True Healing* (Tiburon: HJ Kramer Publishing, 1998), 6-7.

3. J. Rodin and J.E. Langer, "Long-term Effects of Control-relevant Intervention with the Institutionalized Aged," *Journal of Personality and Social Psychology* (1997): vol. 35, no. 12, 897–902.

4. John Robbins, *The Food Revolution: How Your Diet Can Help Save Your Life and Our World* (San Francisco: Conari Press, 2005).

Chapter 17

1. BARF Diet—Healthy & Natural Raw Food for Dogs & Cats, BARF World, Inc., http://www.barfworld.com.

2. Ian Billinghurst, *The Barf Diet: Raw Feeding for Dogs and Cats Using Evolutionary Principles* (N.S.W. Australia: SOS Printing, 2003), 2.

3. Denise Flaim, *The Holistic Dog Book: Canine Care for the 21st Century* (Indianapolis: Wiley and Company, 2003).

4. Jo Stepaniak, *Being Vegan: Living with Conscience, Conviction, and Compassion* (Lincolnwood: Lowell House Publishing, 2000).

5. Denise Flaim, *The Holistic Dog Book: Canine Care for the 21st Century* (Indianapolis: Wiley and Company, 2003).

6. Monica Segal, http://www.monicasegal.com

Index

· ·